ORDINARY DIFFERENTIAL EQUATIONS

ORDINARY DIFFERENTIAL EQUATIONS

by

A. I. Kiselev, M. L. Krasnov

and

G. I. Makarenko

Translated by

E. J. F. Primrose

Senior Lecturer in Mathematics in the University of Leicester

LEICESTER UNIVERSITY PRESS

1967

Printed in Great Britain by Sir Isaac Pitman & Sons Ltd.
at the Pitman Press, Bath,
for
Leicester University Press

CONTENTS

FOREWORD

This collection contains exercises to supplement courses on differential equations in Universities and Technical Colleges.

Particular attention has been give to those questions which have not been explained in sufficient detail in existing textbooks and which, as experiments have shown, are poorly understood by students: for example, a detailed analysis of the method of isoclines, problems on finding orthogonal trajectories, linear dependence and independence of systems of functions.

In spite of its simplicity, the solution of linear differential equations with constant coefficients, homogeneous and non-homogeneous, causes difficulties. Because of this a large number of exercises in the book are devoted to this topic.

Also included in the collection are problems from other sections of a course on differential equations and a certain number of exercises on the solution of boundary value problems, keeping within the general scheme of the solution of linear equations with constant coefficients. It seems to us that students will find it useful to have a knowledge of these.

We have also included problems on Lyapunov stability, on the method of isoclines for equations of the second order, and on the application of the operational method to the solution of differential equations and systems.

At the beginning of each section we give a summary of basic facts and the solution of typical problems. This should help the reader to understand the material.

In compiling the collection of problems we have made use of existing collections, but a considerable proportion of the problems have been compiled by the authors themselves.

Opinions and observations on this collection of problems will be gratefully received.

The authors.

TRANSLATOR'S PREFACE

A few remarks about notations may be helpful. Generally, I have retained the authors' notations. For example, the inverse trigonometrical functions are written arc sin x, arc cos x, etc., though many books give $\sin^{-1} x$, $\cos^{-1} x$, etc. The hyperbolic functions are written sh x, ch x, th x, rather than sinh x, cosh x, tanh x. Polar coordinates are called ρ and ϕ, though r and θ are perhaps more usual. The natural logarithm of x is denoted by ln x, rather than $\log_e x$. I have, however, changed tg x and ctg x to tan x and cot x, to conform with standard English usage.

The original references, which were naturally to the Russian literature, have been replaced by references to works in English, which I trust are nearly equivalent.

I have worked through all the problems and corrected a number of errors in the original; I should like to thank my colleague R. O. Davies for checking all these corrections.

I should like to acknowledge the great help which my wife and daughter gave in preparing the typescript.

E. J. F. Primrose.

§1. BASIC CONCEPTS

A *differential equation* is an equation connecting an independent variable x, an unknown function $y = y(x)$ and its derivatives $y', y'', \ldots, y^{(n)}$, i.e. an equation of the form

$$F(x, y, y', y'', \ldots, y^{(n)}) = 0 \tag{1'}$$

In other words, in a *differential equation* the unknown function appears as a derivative or differential.

If the unknown function $y = y(x)$ is a function of one independent variable, the differential equation is called ordinary. For example:

$$(1)\ \frac{dy}{dx} + xy = 0, \qquad (2)\ y'' + y' + x = \cos x,$$

$$(3)\ (x^2 + y^2)dx - (x + y)dy = 0$$

The *order* of a differential equation is the order of the highest derivative which appears in the equation. For example: $y' + xy = e^x$ is a differential equation of the 1st order, $y'' + p(x)y = 0$ is of the 2nd order, $y^{IX} - xy'' = x^2$ is of the 9th order.

A *solution* of the differential equation (1′) is a function $y = \phi(x)$ which when substituted into the differential equation satisfies it identically. For example, the function $y = \sin x + \cos x$ is a solution of the equation $y'' + y = 0$. For, on differentiating the function twice, we have: $y' = \cos x - \sin x$, $y'' = -\sin x - \cos x$. On substituting the expressions for y'' and y into the differential equation, we obtain the identity $-\sin x - \cos x + \sin x + \cos x \equiv 0$.

The graph of a solution of a differential equation is called an integral curve of the equation.

We shall first consider differential equations of the first order. The general form of an equation of the first order is

$$F(x, y, y') = 0. \tag{1}$$

If the equation (1) can be solved for y', we obtain

$$y' = f(x, y) \tag{2}$$

an equation of the first order, solved for the derivative.

The general solution of the differential equation (2) is a function

$$y = \phi(x, C), \tag{3}$$

depending on one arbitrary constant C and such that:

(*a*) it satisfies equation (2) for any values of the constant C;

(*b*) for any initial condition

$$y|_{x=x_0} = y_0 \; (y(x_0) = y_0) \tag{4}$$

a value $\bar{C}$ of the constant C can be chosen so that the function $y = \phi(x, \bar{C})$ will satisfy the given initial condition (4). A particular solution of the differential equation (2) is a solution obtained from the general solution (3) for some definite value of the arbitrary constant C.

Example. Verify that the function $y = x + C$ is the general solution of the differential equation $y' = 1$, and find the particular solution satisfying the initial condition $y|_{x=0} = 0$. Give a geometrical interpretation of the result.

Solution. The function $y = x + C$ satisfies the given equation for any value of the arbitrary constant C. In fact,

$$y' = (x + C)' = 1.$$

Let us impose an arbitrary initial condition $y|_{x=x_0} = y_0$.

Putting $x = x_0$ and $y = y_0$ into the equation $y = x + C$, we obtain $C = y_0 - x_0$. Substituting this value of C into the given function, we have $y = x + y_0 - x_0$. This function satisfies the given initial condition; in fact, by putting $x = x_0$ we obtain $y = x_0 + y_0 - x_0 = y_0$.

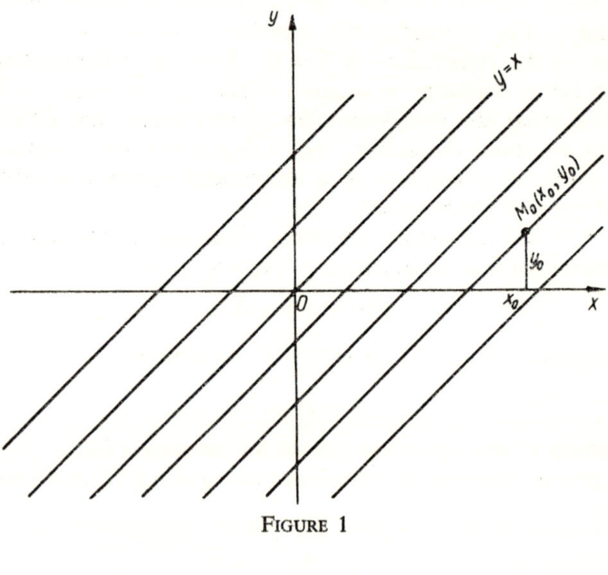

FIGURE 1

So we have shown that the function $y = x + C$ is the general solution of the given equation.

In particular, by putting $x_0 = 0$ and $y_0 = 0$, we obtain the particular solution $y = x$.

The general solution of the given equation, i.e. the function $y = x + C$, defines in the plane XOY a set of parallel lines with gradient $k = 1$. Through each point $M_0(x_0, y_0)$ of the plane XOY there passes the unique integral curve $y = x + y_0 - x_0$. The particular solution $y = x$ defines one of the integral curves, namely the line* which passes through the origin (Fig. 1).

Example. Verify that the function $y = Ce^x$ is the general solution of the equation $y' - y = 0$, and find the particular solution satisfying the initial condition $y|_{x=1} = -1$.

Solution. We have $y = Ce^x$, $y' = Ce^x$. On substituting into the given equation the expressions for y and y', we obtain $Ce^x - Ce^x = 0$, i.e. the function $y = Ce^x$ satisfies the given equation for all values of the constant C.

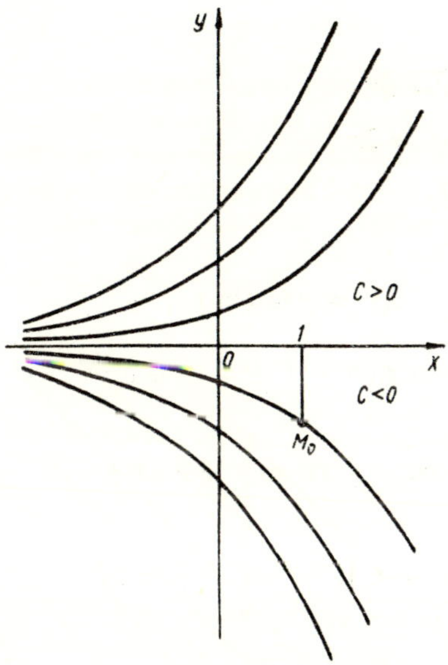

FIGURE 2

Let us impose an arbitrary initial condition $y|_{x=x_0} = y_0$. On substituting x_0 and y_0 for x and y in the function $y = Ce^x$, we have $y_0 = Ce^{x_0}$, so $C = y_0e^{-x_0}$. The function $y = y_0e^{x-x_0}$ satisfies this

* Throughout the book, the term "line" means "straight line."

initial condition. In fact, by putting $x = x_0$, we obtain $y = y_0e^{x_0-x_0} = y_0$. The function $y = Ce^x$ is the general solution of the given equation.

If $x_0 = 1$ and $y_0 = -1$, we obtain the particular solution

$$y = -e^{x-1}.$$

From a geometrical point of view the general solution defines a set of integral curves, which are the graphs of exponential functions; the particular solution is the integral curve which passes through the point $M_0(1, -1)$ (Fig. 2).

In the following problems, show that the given functions are solutions of the differential equations:

1. $y = \frac{\sin x}{x}, \qquad xy' + y = \cos x.$

2. $y = Ce^{-2x} + \frac{1}{3}e^x,$
$y' + 2y = e^x.$

3. $y = 2 + C\sqrt{1 - x^2},$
$(1 - x^2)y' + xy = 2x.$

4. $y = x\sqrt{1 - x^2},$
$yy' = x - 2x^3.$

5. $y = e^{\arcsin Cx}, \qquad xy' = y \tan \ln y.$

6. $y = e^x \int_0^x e^{t^2} dt + Ce^x, \qquad y' - y = e^{x+x^2}.$

7. $y = x \int_0^x \frac{\sin t}{t} dt, \qquad xy' = y + x \sin x.$

8. $y = x\left(\int \frac{e^x}{x} dx + C\right), \qquad xy' - y = xe^x.$

9. $\left.\begin{matrix} x = \cos t \\ y = \sin t \end{matrix}\right\}, \qquad x + yy' = 0.$

10. $\left.\begin{matrix} x = te^t \\ y = e^{-t} \end{matrix}\right\}, \qquad (1 + xy)y' + y^2 = 0.$

11. $\left.\begin{matrix} x = e^{\arctan t} \\ y = e^{-\operatorname{arccot} t} \end{matrix}\right\}, \qquad y - xy' = 0.$

12. $\left.\begin{aligned} x &= t \ln t \\ y &= t^2(2 \ln t + 1) \end{aligned}\right\},\qquad y' \ln \frac{y'}{4} = 4x.$

13. $\left.\begin{aligned} x &= \ln t + \sin t \\ y &= t(1 + \sin t) + \cos t \end{aligned}\right\},\qquad x = \ln y' + \sin y'.$

14. $\left.\begin{aligned} x &= t + \arcsin t \\ y &= \frac{t^2}{2} - \sqrt{1 - t^2} \end{aligned}\right\},\qquad x = y' + \arcsin y'.$

15. $\left.\begin{aligned} x &= t^2 + e^t \\ y &= \tfrac{2}{3}t^3 + (t - 1)e^t \end{aligned}\right\},\qquad y'^2 + e^{y'} = x.$

Show that the functions given in the problems are the general solutions of the differential equations:

16. $y = \dfrac{C}{\cos x}$, $\qquad y' - y \tan x = 0.$

17. $y = -\dfrac{1}{3x + C}$, $\qquad y' = 3y^2.$

18. $y = \ln (C + e^x)$, $\qquad y' = e^{x-y}.$

19. $y = \sqrt{x^2 - Cx}$, $\qquad (x^2 + y^2)\,dx - 2xy\,dy = 0.$

20. $y = x(C - \ln|x|)$, $\qquad (x - y)\,dx + x\,dy = 0.$

21. $x = ye^{Cy+1}$, $\qquad y' = \dfrac{y}{x(\ln x - \ln y)}.$

22. $x = y \ln Cy$, $\qquad y'(x + y) = y.$

A relation of the form $\Phi(x, y, C) = 0$, which defines a general solution implicitly, is a general integral of a differential equation of the first order.

A relation obtained from the general integral for a particular value of the constant C is a particular integral of the differential equation.

The problem of solving or integrating a differential equation consists in finding the general solution or general integral of the given differential equation.

If in addition an initial condition is given, we need to select a particular solution or particular integral which satisfies it.

Since from the geometrical point of view the coordinates x and y

are equivalent, we may consider, instead of the equation $\frac{dy}{dx} = f(x, y)$, the equation $\frac{dx}{dy} = \frac{1}{f(x, y)}$.

In the following problems, verify that the relations are integrals of the differential equations.

23. $e^{-y} - Cx = 1, \qquad xy' + 1 = e^{y}.$

24. $y^3 = \frac{3}{2x} + \frac{C}{x^3}, \qquad xy^2\,dy + y^3\,dx = \frac{dx}{x}.$

25. $x^3 - 4x^2y + 2xy^2 - y^3 = 0,$

$$(3x^2 - 8xy + 2y^2)\,dx - (4x^2 - 4xy + 3y^2)\,dy = 0.$$

26. $y^2 + 2Cx = C^2, \qquad yy'^2 + 2xy' = y.$

27. $\arctan\frac{y}{x} - \ln(C\sqrt{x^2 + y^2}) = 0,$

$$(x + y)\,dx - (x - y)\,dy = 0.$$

28. $x = y\int_0^x \sin t^2\,dt, \qquad y = xy' + y^2 \sin x^2.$

29. $x\int_0^x \frac{\sin t}{t}\,dt = y \ln y, \qquad xy' + xy' \ln y = x \sin x + y \ln y.$

The following theorem concerns the existence and uniqueness of the solution of a differential equation of the first order.

Theorem. Let a differential equation $y' = f(x, y)$ be given, where the function $f(x, y)$ is defined in some region D of the plane XOY containing the point (x_0, y_0). If the function $f(x, y)$ satisfies the following conditions:

(*a*) $f(x, y)$ is a continuous function of the two variables x and y in the region D;

(*b*) $f(x, y)$ has a partial derivative $\frac{\partial f}{\partial y}$ with respect to y, also continuous with respect to x and y in the region D, then there exists a unique solution $y = \phi(x)$ of the given equation satisfying the initial condition $y|_{x=x_0} = y_0$.

The problem of finding the solution of the equation $y' = f(x, y)$ satisfying the initial condition $y|_{x=x_0} = y_0$, is called the Cauchy problem.

Geometrically this means that an integral curve is required which passes through a given point $M_0(x_0, y_0)$ of the plane XOY (Fig. 3).

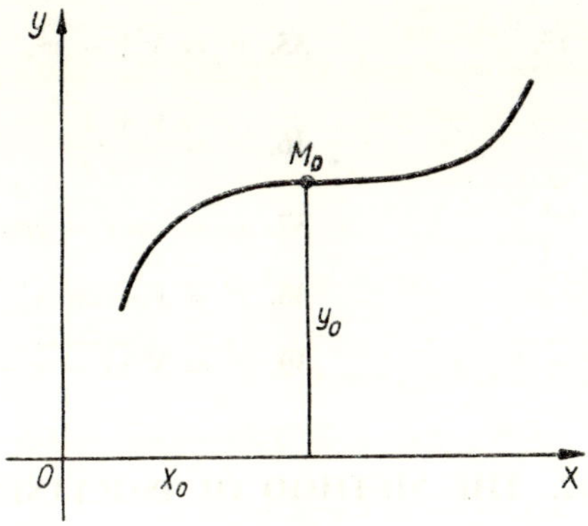

FIGURE 3

Example. Find the region in which the equation $y' = xy + e^{-y}$ has a unique solution.

Solution. The right hand side of the equation $f(x, y) = xy + e^{-y}$ and its partial derivative $\frac{\partial f}{\partial y} = x - e^{-y}$ are continuous with respect to x and y at all points of the plane XOY. By virtue of the theorem of existence and uniqueness, the region in which the given equation has a unique solution is the whole plane XOY.

Example. Find the region in which the equation

$$y' = \tfrac{3}{2}\sqrt[3]{y^2},$$

has a unique solution.

Solution. The right hand side of the equation $f(x, y) = \frac{3}{2}\sqrt[3]{y^2}$ is defined and is continuous at all points of the plane XOY. The partial derivative $\frac{\partial f}{\partial y} = \frac{1}{\sqrt[3]{y}}$ becomes infinite if $y = 0$, i.e. on the axis OX, so if $y = 0$ condition (b) of the theorem of existence and uniqueness is violated. Consequently, at points of the axis OX the uniqueness may break down. It is easy to verify that the function $y = \frac{1}{8}(x + C)^3$ is a solution of the given equation. In addition to this, the equation has an obvious solution $y \equiv 0$. So through any point of the axis OX there pass at least two integral curves, and consequently the uniqueness does break down at points of this axis.

In the following problems, using the theorem of existence and uniqueness, find the regions in which the given equations have a unique solution.

30. $y' = x^2 + y^2$.

31. $y' = \frac{x}{y}$.

32. $y' = y + 3\sqrt[3]{y}$.

33. $y' = \sqrt{x - y}$.

34. $y' = \sqrt{x^2 - y} - x$.

35. $y' = \sqrt{1 - y^2}$.

36. $y' = \frac{y + 1}{x - y}$.

37. $y' = \sin y - \cos x$.

38. $y' = 1 - \cot y$.

39. $y' = \sqrt[3]{3x - y} - 1$.

§2. THE METHOD OF ISOCLINES

The equation

$$y' = f(x, y) \tag{1}$$

defines at each point (x, y), where the function $f(x, y)$ exists, a value of y', i.e. the gradient of the tangent to the integral curve at this point.

If at each point of a region D the value of some quantity is given, we say that the field of this quantity is given in the region D.

Thus the differential equation (1) defines the field of the direction.

The triplet of numbers (x, y, y') defines the direction of a line passing through the point (x, y). The aggregate of segments of these lines gives a geometrical picture of the field of the direction.

The problem of integrating the differential equation (1) can now be interpreted thus: to find a curve such that its tangent at each point has the direction of the field at that point.

The problem of constructing an integral curve can often be solved by introducing isoclines. An *isocline* is the locus of points at which the tangents to the desired integral curves have the same direction. The set of isoclines of the differential equation (1) are defined by by the equation

$$f(x, y) = k, \tag{2}$$

where k is a parameter. By constructing a sufficiently dense net of isoclines, i.e. by giving k closely separated numerical values, we can construct an integral curve of the differential equation (1) sufficiently accurately.

Note 1. The null isocline $f(x, y) = 0$ gives the equation of the curve(s) on which maximum and minimum points of the integral curves lie. For great accuracy in constructing the integral curves we find the locus of the points of inflexion. To do this, we find y'' from equation (1)

$$y'' = \frac{\partial f}{\partial x} + \frac{\partial f}{\partial y} y' = \frac{\partial f}{\partial x} + f(x, y) \frac{\partial f}{\partial y} \tag{3}$$

and equate it to zero. The curve defined by the equation

$$\frac{\partial f}{\partial x} + f(x, y) \frac{\partial f}{\partial y} = 0 \tag{4}$$

is the locus of the points of inflexion, if they exist.

Note 2. The points of intersection of two or more isoclines are singular points of the differential equation (1), since the direction of the integral curve becomes indeterminate there. For example, consider the equation $y' = y/x$. The set of isoclines is defined by the equation $y/x = k$. This is a set of lines, which pass through the origin, i.e. the isoclines, corresponding to the various inclinations

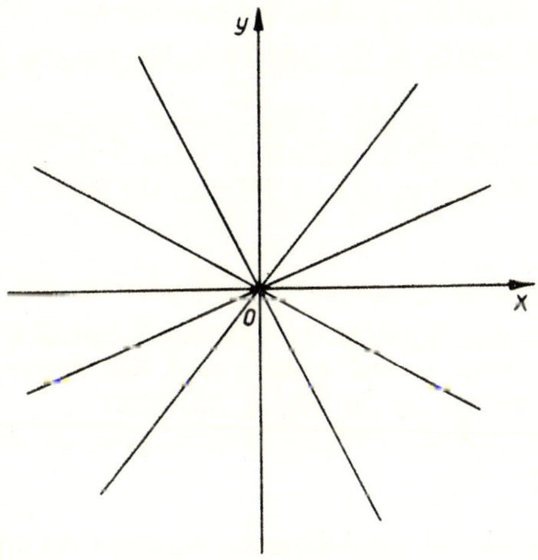

FIGURE 4

of the tangents to the integral curves, intersect at the origin. It is not difficult to verify that the general solution of the given equation has the form $y = Cx$, and the point (0, 0) is a singular point of the differential equation. Here the isoclines are the integral curves of the equation (Fig. 4).

Example. With the help of isoclines, construct approximately the integral curves of the differential equation $y' = 2x - y$.

Solution. To obtain the equation of the isoclines, put $y' = k$ ($k = \text{const.}$). We have

$$2x - y = k \quad \text{or} \quad y = 2x - k.$$

The isoclines are parallel lines. For $k = 0$ we obtain the isocline $y = 2x$. This line divides the plane XOY into two parts, in each of which the derivative y' always has the same sign (Fig. 5).

The integral curves which intersect the line $y = 2x$ pass from the region where the function y decreases to the region where it increases, and vice versa, and this means that the turning points of the integral curves, in fact the minima, lie on this line.

We now take the two isoclines

$$k = -1, \quad y = 2x + 1 \quad \text{and} \quad k = 1, \quad y = 2x - 1.$$

The tangents drawn to the integral curves at the points of intersection with the isoclines $k = -1$ and $k = 1$ make angles of 135° and 45° with the axis OX. We also find the second derivative $y'' = 2 - y' = 2 - 2x + y$.

The line $y = 2x - 2$, on which $y'' = 0$, is the isocline obtained for $k = 2$, and at the same time an integral curve, which can be verified by substituting into the equation. Since the right hand side of the given equation $f(x, y) = 2x - y$ satisfies the conditions of the theorem of existence and uniqueness over the whole plane XOY, the other integral curves do not intersect this isocline. The isocline $y = 2x$, on which the minimum points of the integral curves lie, is situated above the isocline $y = 2x - 2$, and so the integral curves which lie below the isocline $y = 2x - 2$ do not have turning points.

The line $y = 2x - 2$ divides the plane XOY into two parts, in one of which (situated above the line) $y'' > 0$, which means that the integral curves are concave upwards, and in the other $y'' < 0$, which means that the integral curves are concave downwards. Since the integral curves do not intersect the line $y = 2x - 2$, it is not the locus of points of inflexion. The integral curves of the given equation do not have points of inflexion.

This analysis enables us to construct approximately the set of integral curves of the equation (Fig. 5).

Example. By the method of isoclines construct approximately the integral curves of the differential equation $y' = \sin(x + y)$.

Solution. By putting $y' = k$, where $k = \text{const.}$, we obtain the equation of the isoclines $\sin(x + y) = k$, where $-1 \leqslant k \leqslant 1$.

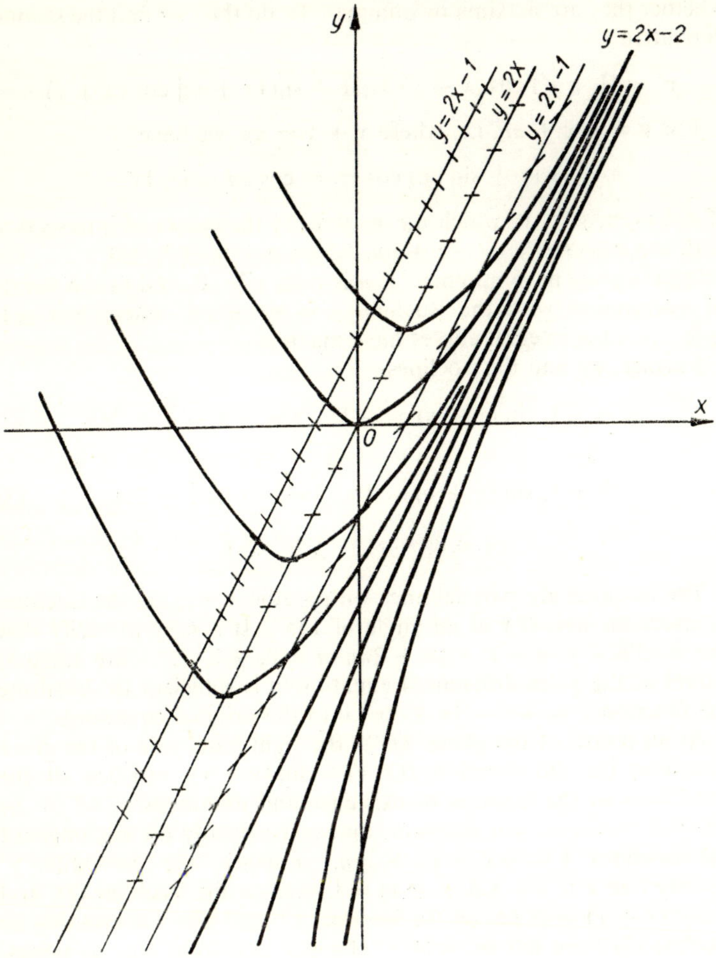

FIGURE 5

If $k = 0$,

$$\sin(x + y) = 0, \quad \text{hence} \quad y = -x + \pi n \qquad (1)$$

$$(n = 0, \quad \pm 1, \quad \pm 2, \ldots).$$

The integral curves have horizontal tangents at their points of intersection with these isoclines. We determine whether the integral curves have turning points on the isoclines $y = -x + \pi n$, and if so, whether they are maxima or minima. To do this, we find the second derivative

$$y'' = (1 + y') \cos (x + y) = [1 + \sin (x + y)] \cos (x + y).$$

For $y = -x + \pi n$, i.e. where $x + y = \pi n$, we have

$$y'' = (1 + \sin \pi n) \cos \pi n = \cos \pi n = (-1)^n.$$

If n is even, $y'' > 0$, which means that, at the points of intersection with the isoclines $y = -x + \pi n$, where $n = 0, \pm 2, \pm 4, \ldots$, the integral curves have minima; if n is odd, $y'' < 0$, and at the points of intersection with the isoclines $y = -x + \pi n$, where $n = \pm 1, \pm 3, \ldots$, the integral curves have maxima.

Further, we find the isoclines

$$k = -1, \sin (x + y) = -1; \quad y = -x - \frac{\pi}{2} + 2\pi n \tag{2}$$

$$k = 1, \sin (x + y) = 1; \quad y = -x + \frac{\pi}{2} + 2\pi n \tag{3}$$

$$(n = 0, \quad \pm 1, \quad \pm 2, \ldots).$$

The isoclines are parallel lines with gradient -1, i.e. the isoclines intersect the axis OX at an angle of 135°. It is easy to verify that the isoclines $y = -x - \frac{1}{2}\pi + 2\pi n$ ($n = 0, \pm 1, \ldots$) are integral curves of the given differential equation (it is sufficient to substitute the function $y = -x - \frac{1}{2}\pi + 2\pi n$ into the original equation).

At all points of the plane XOY, the right hand side of the given equation, i.e. the function $f(x, y) = \sin (x + y)$, satisfies all the conditions of the theorem of existence and uniqueness, and so the integral curves do not intersect, and consequently do not intersect the isoclines $y = -x - \frac{1}{2}\pi + 2\pi n$. Further, the derivative y'' becomes zero if $1 + \sin (x + y) = 0$, i.e. on the isoclines (2), and if $\cos (x + y) = 0$, i.e. on the isoclines (2) and (3). On crossing an isocline (3) from left to right y'' changes sign from plus to minus. For example, if we consider the region enclosed between the isoclines $y = -x$ and $y = -x + \pi$, then on the isocline $y = -x + \frac{1}{2}\pi$ we have $y'' = 0$; moreover, below the isocline $y'' > 0$, which means that the integral curves are concave upwards, and above the isocline $y'' < 0$, which means that the integral curves are concave downwards. Thus the isoclines (3) are the locus of the points of inflexion of the integral curves. The data obtained enable us to construct approximately the set of integral curves of the given equation. For a more accurate construction, one must construct still more isoclines (Fig. 6).

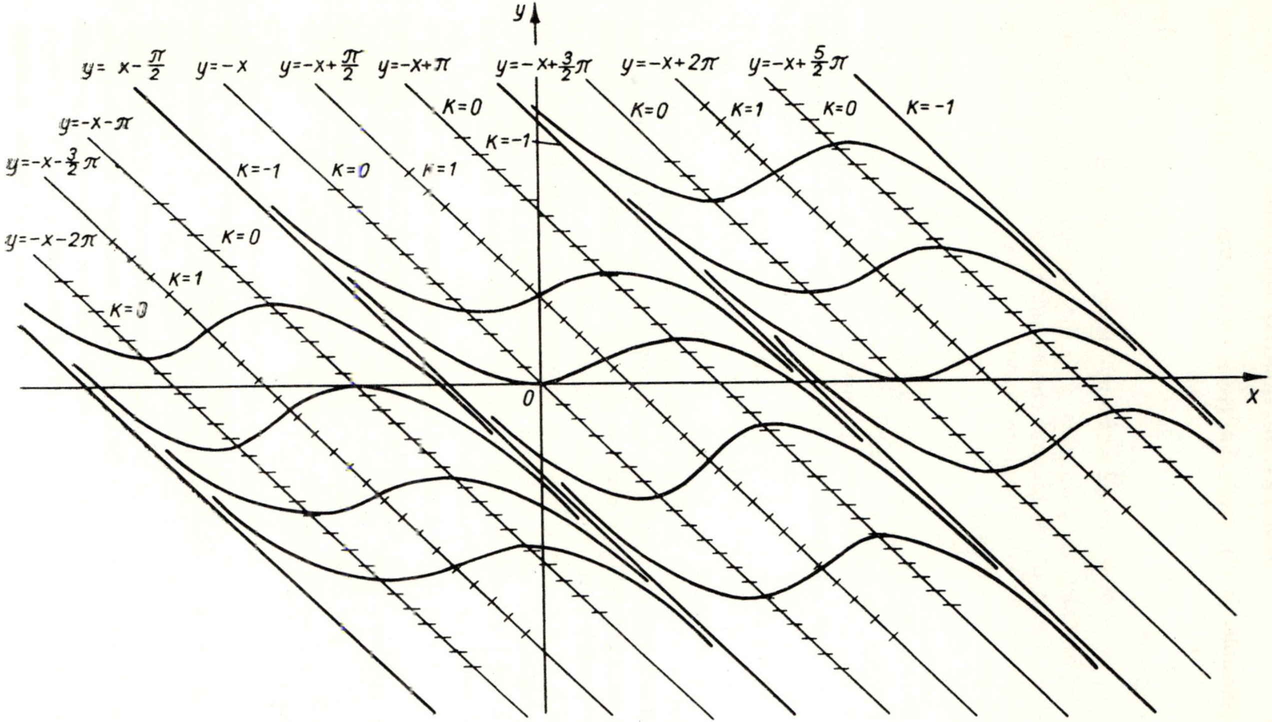
y
y= x- π/2
y=-x
y=-x+ π/2
y=-x+π
y=-x+ 3/2π
y=-x+2π
y=-x+ 5/2π
y=-x-π
y=-x- 3/2π
y=-x-2π
K=0
K=-1
K=1
K=0
K=1
K=0
K=-1
K=-1
K=0
K=1
K=0
K=1
K=0
0
x

FIGURE 6

Example. By the method of isoclines, construct the integral curves of the equation $y' = y - x^2 + 2x - 2$.

Solution. Put $y' = k$ (k = const.). Then the equation of the isoclines is

$$y - x^2 + 2x - 2 = k \quad \text{or} \quad y = x^2 - 2x + 2 + k.$$

The isoclines are parabolas with the vertical axis of symmetry $x = 1$. There are no integral curves among the isoclines. In fact, by substituting into the given equation $y = x^2 - 2x + 2 + k$ and $y' = 2x - 2$, we have

$$2x - 2 = x^2 - 2x + 2 + k - x^2 + 2x - 2 \quad \text{or} \quad 2x - 2 = k.$$

But this equation cannot be satisfied identically for any value of k.

Put $k = 0$. Then at the points of intersection with the isocline $y = x^2 - 2x + 2$ the integral curves will have horizontal tangents. The isocline $k = 0$, i.e. the parabola $y = x^2 - 2x + 2$ divides the plane XOY into two parts: in one of them $y' < 0$ (y decreasing) and in the other $y' > 0$ (y increasing). Since this isocline is not an integral curve, the turning points of the integral curves lie on it, namely maxima on the part of the parabola $y = x^2 - 2x + 2$ where $x > 1$, and minima on the part of the parabola where $x < 1$. The integral curve which passes through the point (1, 1), i.e. through the vertex of the parabola $y = x^2 - 2x + 2$, does not have a turning point there. At points of the isoclines $k = 1$, $y = x^2 - 2x + 3$ and $k = -1$, $y = x^2 - 2x + 1$, the tangents to the integral curves have gradients equal to 1 and -1 respectively.

To find the direction of concavity of the integral curves, we find the second derivative. We have

$$y'' = y' - 2x + 2 = y - x^2 + 2x - 2 - 2x + 2 = y - x^2.$$

It is zero only at points lying on the parabola $y = x^2$. At points of the plane XOY whose coordinates satisfy the condition $y < x^2$, the integral curves are concave downwards ($y'' < 0$), and at points where $y > x^2$ they are concave upwards ($y'' > 0$). The points of intersection of the integral curves with the parabola $y = x^2$ are points of inflexion of these curves. So the parabola $y = x^2$ is the locus of points of inflexion of the integral curves.

The right hand side of the original equation

$$f(x, y) = y - x^2 + 2x - 2$$

satisfies the conditions of the theorem of existence and uniqueness at all points of the plane XOY, so a unique integral curve of the equation passes through each point of the plane.

By using the information obtained, we may construct approximately the set of integral curves of the given equation (Fig. 7).

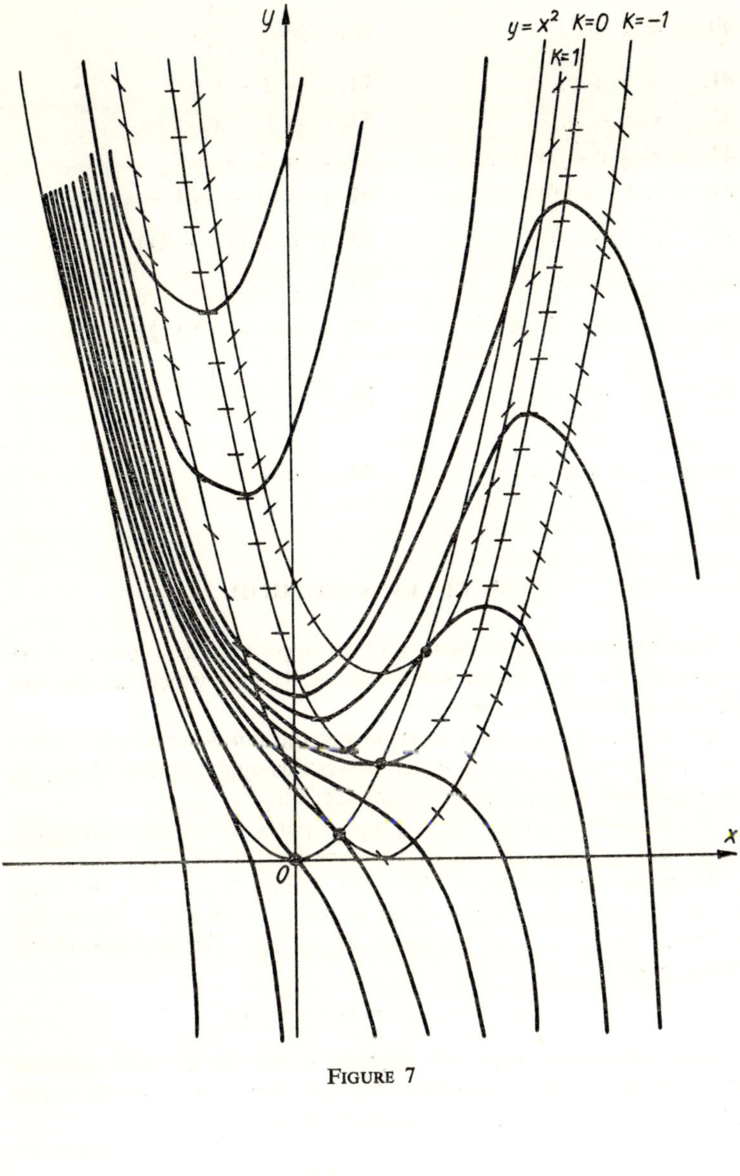

FIGURE 7

By the method of isoclines, construct the integral curves of the following differential equations:

40. $y' = x + 1$.

41. $y' = x + y$.

42. $y' = y - x$.

43. $y' = \frac{1}{2}(x - 2y + 3)$.

44. $y' = (y - 1)^2$.

45. $y' = (y - 1)x$.

46. $y' = x^2 + y^2$.

47. $y' = \cos(x - y)$.

48. $y' = y - x^2$.

49. $y' = x^2 + 2x - y$.

50. $y' = \dfrac{y + 1}{x - 1}$.

51. $y' = 2 - y$.

52. $y' = 1 - x$.

53. $y' = 2x - y$.

54. $y' = (1 - y)(1 - x)$.

55. $y' = \sin(y - 2x)$.

56. $y' = x^2 + y$.

57. $y' = y - x^2 + 2x$.

58. $y' = \dfrac{x - 1}{y}$.

59. $y' = -\dfrac{y}{x}$.

§3. EULER'S METHOD

This method enables us to find an approximate solution of the equation $y' = f(x, y)$ on some interval $[x_0, b]$, which satisfies the initial condition $y|_{x=x_0} = y_0$.

Euler's method consists in replacing the unknown integral curve of the differential equation which passes through the point $M_0(x_0, y_0)$ by a polygon consisting of straight segments.

We divide the interval $[x_0, b]$ into n parts, not necessarily equal, by points $x_0 < x_1 < x_2 < \ldots < x_n = b$.

From the starting point $M_0(x_0, y_0)$ of the integral curve we draw a line M_0M_1, with gradient equal to $f(x_0, y_0)$, up to its intersection with the line $x = x_1$ at the point $M_1(x_1, y_1)$. The ordinate of the point M_1 is

$$y_1 = y_0 + f(x_0, y_0)(x_1 - x_0).$$

From the point $M_1(x_1, y_1)$ we draw a line M_1M_2, with gradient $f(x_1, y_1)$, up to its intersection with the line $x = x_2$ at the point $M_2(x_2, y_2)$. The ordinate of the point M_2 is

$$y_2 = y_1 + f(x_1, y_1)(x_2 - x_1).$$

Similarly we define the point $M_3(x_3, y_3)$ and so on. The ordinate of the point $M_n(x_n, y_n)$ is

$$y_n = y_{n-1} + f(x_{n-1}, y_{n-1})(x_n - x_{n-1}).$$

The quantities $y_1, y_2, \ldots, y_n$ are approximate values of the solution of the given equation at the points $x_1, x_2, \ldots, x_n$.

As a result of this construction we obtain a polygon called the Euler polygon, which represents approximately the integral curve which passes through the point $M_0(x_0, y_0)$ (Fig. 8).

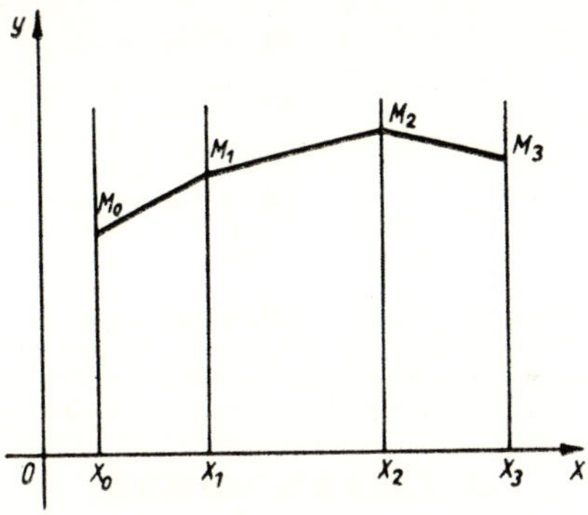

FIGURE 8

Usually for ease of calculation and estimation the interval $[x_0, b]$ is divided into equal parts and $x_k - x_{k-1}$ is denoted by h. The quantity h is called a step of the independent variable.

It can be shown that with certain conditions on the function $f(x, y)$ the approximate solution gives, as $h \to 0$, an exact solution of the given equation, satisfying the initial condition $y|_{x=x_0} = y_0$ (see R. A. Struble, *Non-Linear Differential Equations*, Chapter 2).

Example. Using the method of Euler polygons, find on the interval [0, 1] an approximate solution of the equation $y' = 2x - y$, satisfying the initial condition $y|_{x=0} = -1$. Divide the interval [0, 1] into 10 equal parts and compare the approximate values of the solution at the points of division with the values of the exact solution $y = 2x - 2 + e^{-x}$ at these points.

Solution. The values of the ordinates y_k of the points $M_k(x_k, y_k)$ are calculated from the formula $y_k = y_{k-1} + f(x_{k-1}, y_{k-1})(x_k - x_{k-1})$

$$(k = 1, 2, 3, 4, 5, 6, 7, 8, 9, 10).$$

In the given example, $f(x_{k-1}, y_{k-1}) = 2x_{k-1} - y_{k-1}$ and the difference $x_1 - x_0 = x_2 - x_1 = \ldots = x_{10} - x_9 = h = 0{\cdot}1$, since the interval [0, 1] is divided into 10 equal parts, and consequently

$$y_k = y_{k-1} + 2(x_{k-1} - y_{k-1}) \times 0{\cdot}1, \text{ where } k = 1, 2, \ldots, 10.$$

We reduce all calculations to a table, in the last column of which we put the values of the exact solution $y = 2(x - 1) + e^{-x}$ at the points of division of the interval [0, 1].

k	x_k	$2x_k$	y_k	$2x_k - y_k$	$(2x_k - y_k)\,0{\cdot}1$	y_{k+1}	$y_k = 2x_k - 2 + e^{x}k$
0	0	0	−1	1	0·1	−0·900	−1
1	0·1	0·2	−0·9000	1·1	0·1100	−0·7900	−0·8952
2	0·2	0·4	−0·7900	1·19	0·1190	−0·6710	−0·7813
3	0·3	0·6	−0·6710	1·271	0·1271	−0·5439	−0·6592
4	0·4	0·8	−0·5439	1·3439	0·13439	−0·4095	−0·5297
5	0·5	1·0	−0·4095	1·4095	0·14095	−0·2686	−0·3935
6	0·6	1·2	−0·2686	1·4686	0·14686	−0·1217	−0·2512
7	0·7	1·4	−0·1217	1·5217	0·15217	0·03047	−0·1034
8	0·8	1·6	−0·03047	1·5695	0·15695	0·1874	0·0493
9	0·9	1·8	−0·1874	1·6126	0·16126	0·3486	0·2066
10	1·0	2·0	−0·3486				0·3679

We see that Euler's method does not give great accuracy, but in certain cases satisfactory accuracy. For example, for $x = 0{\cdot}5$ the absolute error in the approximate value of the solution is

$$\Delta = |-0{\cdot}4095 + 0{\cdot}3935| = 0{\cdot}0160;$$

and the relative error is

$$\delta = \frac{0{\cdot}0160}{|-0{\cdot}4095|} = 0{\cdot}039 \approx 4\%.$$

Solve the following problems by Euler's method and the use of a slide rule.

60. Find the value for $x = 1$ of that solution of the equation $y' = x^2y + 2$ which satisfies the initial condition $y|_{x=0} = 0$ $(h = 0{\cdot}2)$.

61. Construct approximately on the interval [1, 3] the integral curve of the equation $y' = x + y$ which passes through the point $M(1, 2)$, and calculate $y(3)$ $(h = 0{\cdot}2)$.

For the following equations, set up, on the given interval, a table of values of the solution which satisfies the given initial condition:

62. $y' = \dfrac{y}{x}$, $y(1) = 1$, $[1, 4]$ $(h = 0{\cdot}5)$.

63. $y' = \frac{1}{2}xy$, $y(0) = 1$, $[0, 1]$ $(h = 0{\cdot}1)$.

64. $y' = x^2 + y^2$, $y(0) = 0$, $[0, 1]$ $(h = 0{\cdot}1)$.

65. $y' = 1 + xy^2$, $y(0) = 0$, $[0, 1]$ $(h = 0{\cdot}1)$.

66. $y' = \dfrac{y}{x+1} - y^2$, $y(0) = 1$, $[0, 1]$ $(h = 0{\cdot}1)$.

§4. EQUATIONS WITH SEPARABLE VARIABLES

A differential equation of the form

$$\phi(y)\,dy = f(x)\,dx \tag{1}$$

is called an equation with separated variables.

An equation of the form

$$\phi_1(x)\psi_1(y)\,dx = \phi_2(x)\psi_2(y)\,dy, \tag{2}$$

in which the coefficients of the differentials split up into factors depending only on x and only on y is called an equation with separable variables.

On division by the product $\psi_1(y)\phi_2(x)$ it reduces to an equation with separated variables:

$$\frac{\phi_1(x)}{\phi_2(x)}\,dx = \frac{\psi_2(y)}{\psi_1(y)}\,dy. \tag{2'}$$

The general integral of this equation has the form

$$\int \frac{\phi_1(x)}{\phi_2(x)}\,dx - \int \frac{\psi_2(y)}{\psi_1(y)}\,dy = C. \tag{3}$$

Note. Division by $\psi_1(y)\phi_2(x)$ can lead to the loss of particular solutions, obtained by equating to zero the product $\psi_1(y)\phi_2(x)$.

A differential equation of the form

$$\frac{dy}{dx} = f(ax + by + c),$$

where a, b and c are constants, may be transformed into an equation with separable variables by means of a change of variables

$$z = ax + by + c.$$

Example. Solve the equation

$$3e^x \tan y \, dx + (2 - e^x) \sec^2 y \, dy = 0.$$

Solution. We divide both sides of the equation by the product $(2 - e^x)\tan y$:

$$\frac{3e^x \, dx}{2 - e^x} + \frac{\sec^2 y \, dy}{\tan y} = 0.$$

This equation has separated variables. We find its integral

$$-3 \ln |2 - e^x| + \ln |\tan y| = C_1.$$

Taking the exponential of each side we obtain

$$\frac{|\tan y|}{|2 - e^x|^3} = e^{C_1} \quad \text{or} \quad \left|\frac{\tan y}{(2 - e^x)^3}\right| = e^{C_1}.$$

It follows that

$$\frac{\tan y}{(2 - e^x)^3} = \pm e^{C_1}.$$

Writing $\pm e^{C_1} = C$, we have

$$\frac{\tan y}{(2 - e^x)^3} = C \quad \text{or} \quad \tan y - C(2 - e^x)^3 = 0.$$

We have obtained the general integral of the given equation.

On dividing by the product $(2 - e^x) \tan y$, we have assumed that neither factor becomes zero. Equating each factor to zero, we obtain respectively

$$y = k\pi \; (k = 0, \pm 1, \pm 2, \ldots), \; x = \ln 2.$$

By direct substitution into the original equation we verify that $y = k\pi$ and $x = \ln 2$ are solutions of the equation. They may be obtained formally from the general integral by putting $C = 0$ and $C = \infty$. The latter denotes that the constant C is replaced by $1/C_2$, so that the general integral takes the form

$$\tan y - \frac{1}{C_2}(2 - e^x)^3 = 0 \quad \text{or} \quad C_2 \tan y - (2 - e^x)^3 = 0;$$

by putting $C_2 = 0$ in the latter equation, which corresponds to $C = \infty$, we have $(2 - e^x)^3 = 0$, from which we obtain the solution $x = \ln 2$ of the original equation. Thus, the functions

$$y = k\pi \ (k = 0, \pm 1, \pm 2, \ldots) \quad \text{and} \quad x = \ln 2$$

are particular solutions of the given equation. So the final answer is

$$\tan y - C(2 - e^x)^3 = 0.$$

Example. Find the particular solution of the equation

$$(1 + e^x)yy' = e^x,$$

satisfying the initial condition $y|_{x=0} = 1$.

Solution. We have

$$(1 + e^x)y\frac{dy}{dx} = e^x.$$

Separating the variables, we obtain

$$y\,dy = \frac{e^x\,dx}{1 + e^x}.$$

Integrating, we find the general integral

$$\frac{y^2}{2} = \ln(1 + e^x) + C. \tag{1}$$

Putting $x = 0$, $y = 1$ in (1), we have $\frac{1}{2} = \ln 2 + C$; hence we find that $C = \frac{1}{2} - \ln 2 = \ln(\frac{1}{2}\sqrt{e})$. Substituting in (1) we obtain the particular integral

$$\tfrac{1}{2}y^2 = \ln(1 + e^x) + \ln\frac{\sqrt{e}}{2}.$$

Taking the exponential of each side, we have

$$e^{\frac{y^2}{2}} = \frac{\sqrt{e}}{2}(1 + e^x).$$

Example. Find the particular solution of the equation

$$y' \sin x = y \ln y,$$

satisfying the initial conditions

$$\text{(a) } y\Big|_{x=\frac{\pi}{2}} = e, \qquad \text{(b) } y\Big|_{x=\frac{\pi}{2}} = 1.$$

Solution. We have

$$\frac{dy}{dx}\sin x = y \ln y.$$

We separate the variables:

$$\frac{dy}{y \ln y} = \frac{dx}{\sin x}.$$

Integrating, we find the general integral

$$\ln \left| \ln y \right| = \ln \left| \tan \frac{x}{2} \right| + \ln C.$$

Taking the exponential of each side, we obtain

$$\ln y = C \tan \frac{x}{2} \quad \text{or} \quad y = e^{C \tan \frac{x}{2}},$$

which is the general solution of the original equation.

(a) We now put $x = \frac{1}{2}\pi$, $y = e$, then $e = e^{C \tan \frac{1}{4}\pi}$. Hence $C = 1$. So $y = e^{\tan \frac{1}{2}x}$.

(b) We now find the particular solution of the equation which satisfies the initial condition $y|_{x=\frac{1}{2}\pi} = 1$.

From the general solution $y = e^{C \tan \frac{1}{2}x}$ we obtain, for $C = 0$, the required particular solution. We note that in the process of obtaining the general solution the constant C entered under the logarithm sign, and that means that $C = 0$ must be regarded as a limiting value. This particular solution $y = 1$ is contained among the zeros of the product $y \ln y \sin x$, by which we divided both sides of the given equation.

Integrate the equations:

67. $(1 + y^2)\, dx + (1 + x^2)\, dy = 0.$

68. $(1 + y^2)\, dx + xy\, dy = 0.$

69. $(y^2 + xy^2)y' + x^2 - yx^2 = 0.$

70. $(1 + y^2)\, dx = x\, dy.$

71. $x\sqrt{1 + y^2} + yy'\sqrt{1 + x^2} = 0.$

72. $x\sqrt{1 - y^2}\, dx + y\sqrt{1 - x^2}\, dy = 0, \qquad y|_{x=0} = 1.$

73. $e^{-y}(1 + y') = 1.$

74. $y \ln y\, dx + x\, dy = 0, \qquad y|_{x=1} = 1.$

75. $y' = a^{x+y} (a > 0, a \neq 1)$.

76. $e^y(1 + x^2)\,dy - 2x(1 + e^y)\,dx = 0$.

77. $(1 + e^x)yy' = e^y, \qquad y|_{x=0} = 0$.

78. $(1 + y^2)(e^{2x}\,dx - e^y\,dy) - (1 + y)\,dy = 0$.

79. $(xy^2 - y^2 + x - 1)\,dx + (x^2y - 2xy + x^2 + 2y - 2x + 2)\,dy = 0$.

80. $y' = \sin(x - y)$.

81. $y' = ax + by + c \qquad (a, b, c - \text{const.})$.

82. $(x + y)^2y' = a^2$.

83. $(x - y^2)\,dx + 2xy\,dy = 0$.

84. $(1 + y^2)\,dx = (y - \sqrt{1 + y^2})(1 + x^2)^{3/2}\,dy$.

85. $xy^2(xy' + y) = a^2$.

86. $(x^2y^2 + 1)\,dx + 2x^2\,dy = 0$
(substitute $xy = t$).

87. $(x^3y^3 + x^2y^2 + xy + 1)y + (x^3y^3 - x^2y^2 - xy + 1)xy' = 0$
(substitute $xy = t$).

88. $(x^2y^3 + y + x - 2)\,dx + (x^3y^2 + x)\,dy = 0$
(substitute $xy = t$).

89. $(x^6 - 2x^5 + 2x^4 - y^3 + 4x^2y)\,dx + (xy^2 - 4x^3)\,dy = 0$
(substitute $y = tx$).

90. $y' + 1 = \dfrac{(x + y)^m}{(x + y)^n + (x + y)^p}$.

91. $(\ln x + y^3)\,dx - 3xy^2\,dy = 0$.

92. $(xy + 2xy\ln^2 y + y\ln y)\,dx + (2x^2\ln y + x)\,dy = 0$
(substitute $x\ln y = t$).

93. $y - xy' = a(1 + x^2y')$.

94. $(a^2 + y^2)\,dx + 2x\sqrt{ax - x^2}\,dy = 0, \qquad y|_{x=a} = 0$.

95. $y' + \sin\dfrac{x + y}{2} = \sin\dfrac{x - y}{2}$.

96. Find the curve passing through the point (0, −2) such that the gradient of the tangent at any point is equal to the ordinate of that point, multiplied by three units.

97. Find the curve such that the area Q enclosed by the curve, the axis OX and the two ordinates $X = 0$, $X = x$ is a given function of y:

$$Q = a^2 \ln \frac{y}{a}.$$

98. A particle of mass 1 gram moves along a straight line under the action of a force which is directly proportional to the time, measured from the instant $t = 0$, and inversely proportional to the speed of the particle. When $t = 10$ sec the speed is 50 cm/sec and the force is 4 dynes. What is the speed a minute after the start of the motion?

99. Prove that a curve defined by the property that all its normals pass through a fixed point is a circle.

100. A bullet enters a block of thickness $h = 10$ cm with speed $v_0 = 200$ m/sec and leaves the block, having passed through it, with speed $v_1 = 80$ m/sec. Assuming that the resistance of the block to the motion of the bullet is proportional to the square of the speed, find the time taken for the bullet to go through the block.

101. A ship is retarded by the resistance of the water, which is proportional to the speed of the ship. The initial speed of the ship is 10 m/sec, and the speed after 5 sec is 8 m/sec. When will the speed decrease to 1 m/sec?

102. Prove that a curve such that the gradient of the tangent at any point is proportional to the abscissa of the point of contact is a parabola.

103. According to Newton's law, the rate of cooling of a body in air is proportional to the difference between the temperature T of the body and the temperature T_0 of the air. If the temperature of the air is 20°C and the body cools from 100° to 60° in 20 min., in what time will the temperature fall to 30°?

104. Find a curve such that the gradient of the tangent at any point is n times the gradient of the line joining that point to the origin.

105. Determine the distance s described by a body in time t if its speed is proportional to the distance described and if the body describes 100 m. in 10 seconds and 200 m. in 15 seconds.

106. The bottom of a reservoir of capacity 300 litres is covered with a mixture of salt and insoluble material. Assuming that the rate of dissolution of the salt is proportional to the difference between the concentration at the given moment and the concentration of the saturated solution (1 kg of salt in 3 litres of water) and that the given quantity of pure water dissolves $\frac{1}{3}$ kg of salt in one minute, find how much salt the solution will contain after one hour.

107. A certain quantity of porous insoluble material contains 10 kg of salt. It is found that, under the action of 90 litres of water, half the salt contained in it is dissolved in one hour. How much salt would have been dissolved in the same time if the quantity of water had been doubled? The rate of dissolution is proportional to the quantity of undissolved salt and to the difference between the concentration of solution at the given moment and the concentration of the saturated solution (1 kg in 3 litres).

108. The quantity of light absorbed on passing through a thin layer of water is proportional to the thickness of the layer and to the quantity of light falling on its surface. If half the original quantity of light is absorbed on passing through a layer of thickness 3 m, what fraction will pass through a depth of 30 m?

109. A certain quantity of material containing 3 kg of moisture is placed in a room of capacity 100 m³ in which the air has originally 25% humidity. Saturated air at the same temperature contains 0·12 kg of moisture per cubic metre. If during the first day the material loses half its moisture, how much of its moisture will remain at the end of the second day?

Instructions. The moisture contained in porous material evaporates into the surrounding space at a rate proportional to the quantity of moisture in the given material, and also proportional to the difference between the humidity of the surrounding air and the humidity of saturated air.

110. A certain quantity of porous insoluble material, containing 2 kg of salt, is subjected to the action of 30 litres of water. In 5 minutes, 1 kg of salt is dissolved. In what time will 99% of the original quantity of salt be dissolved?

111. A brick wall ($k = 0{\cdot}0015$) has a thickness of 30 cm. Find the relation between the temperature and the distance from the outer surface, if the temperature is 20° at the inner and 0° at the outer surface of the wall. Find also the quantity of heat which the wall gives out (per square metre) in the course of a day.

Instructions. By virtue of Newton's law, the rate Q at which heat is conducted across an area perpendicular to the x-axis is equal to $Q = -kS\,dT/dx$, where k is the coefficient of heat conductivity of the given material, T is the temperature, and S is the measure of the area.

112. Show that the equation $\dfrac{dy}{dx} = \dfrac{y}{x}$ with the initial condition $y|_{x=0} = 0$ has an infinite set of solutions of the form $y = Cx$. The same equation with the initial condition $y|_{x=0} = y_0 \neq 0$ has no solution. Construct the integral curves.

113. Show that the problem

$$\frac{dy}{dx} = y^{\alpha}, \quad y|_{x=0} = 0$$

has at least two solutions if $0 < \alpha < 1$ and one solution if $\alpha = 1$. Construct the integral curves for $\alpha = \frac{1}{2}$, 1.

114. Find a solution of the equation

$$\frac{dy}{dx} = y(\ln y)^{\alpha} \ (\alpha > 0),$$

satisfying the initial condition $y|_{x=0} = 0$. For what values of α has it a unique solution?

115. Show that the tangents to all integral curves of the differential equation

$$y' + y \tan x = x \tan x + 1$$

at their points of intersection with the axis OY are parallel. Determine the angle at which the integral curves cut the axis OY.

Integrate the following differential equations:

116. $\cos y' = 0$ (draw a figure).

117. $e^{y'} = 1$.

118. $\sin y' = x$.

119. $\ln y' = x$.

120. $\tan y' = 0$ (draw a figure).

121. $e^{y'} = x$.

122. $\tan y' = x$.

In the following problems, find the solutions of the equations satisfying the given conditions as $x \to \pm\infty$:

123. $x^2 \cos y \, y' + 1 = 0$, $\quad y \to \frac{16}{3}\pi$, $x \to +\infty$.

124. $x^2 y' + \cos 2y = 1$, $\quad y \to \frac{10}{3}\pi$, $x \to +\infty$.

125. $x^3 y' - \sin y = 1$, $\quad y \to 5\pi$, $x \to +\infty$.

126. $2(1 + x^2)y' - \cos^2 2y = 0, \qquad y \to \frac{7}{2}\pi$
$x \to -\infty.$

127. $e^y = e^{4y}y' + 1, \qquad y$ bounded as $x \to +\infty.$

128. $(x + 1)y' = y - 1, \qquad y$ bounded as $x \to +\infty.$

129. $y' = 2x(\pi + y), \qquad y$ bounded as $x \to \infty.$

130. $x^2y' + \sin 2y = 1, \qquad y \to \frac{11}{4}\pi$
$x \to +\infty.$

§5. HOMOGENEOUS EQUATIONS

A function $f(x, y)$ is a homogeneous function of its arguments, of degree n, if

$$f(tx, ty) \equiv t^n f(x, y).$$

For example, the function $f(x, y) = x^2 + y^2 - xy$ is a homogeneous function of degree 2, since $f(tx, ty) = (tx)^2 + (ty)^2 - (tx)(ty) = t^2(x^2 + y^2 - xy) = t^2f(x, y)$.

If $n = 0$ we have a function of degree zero. For example,

$$f(x, y) = \frac{x^2 - y^2}{x^2 + y^2}$$

is a homogeneous function of degree zero, since

$$f(tx, ty) = \frac{(tx)^2 - (ty)^2}{(tx)^2 + (ty)^2} = \frac{t^2(x^2 - y^2)}{t^2(x^2 + y^2)} = \frac{x^2 - y^2}{x^2 + y^2} = f(x, y).$$

A differential equation of the form $\dfrac{dy}{dx} = f(x, y)$ is called homogeneous if $f(x, y)$ is a homogeneous function of its arguments, of degree zero. A homogeneous equation can always be expressed in the form

$$\frac{dy}{dx} = \phi\left(\frac{y}{x}\right). \tag{1}$$

If we introduce a new unknown function $u = y/x$, equation (1) leads to an equation with separable variables:

$$x\frac{du}{dx} = \phi(u) - u.$$

If the differential equation is written in the form

$$P(x, y)\,dx + Q(x, y)\,dy = 0$$

(equation in differentials), it is homogeneous if $P(x, y)$ and $Q(x, y)$ are homogeneous functions of the same degree.

Let $u = u_0$ be a root of the equation $\phi(u) - u = 0$. Then a solution of the homogeneous equation is $u = u_0$ or $y = u_0x$, a line passing through the origin.

Note 1. To solve a homogeneous equation it is not essential to reduce it to the form (1). The substitution $y = ux$ can be used at once.

An equation of the form

$$\frac{dy}{dx} = f\left(\frac{a_1x + b_1y + c_1}{a_2x + b_2y + c_2}\right) \tag{2}$$

can be transformed into a homogeneous equation by means of a change of origin to the point of intersection (x_0, y_0) of the lines

$$a_1x + b_1y + c_1 = 0 \quad \text{and} \quad a_2x + b_2y + c_2 = 0.$$

This is done by a change of variables

$$x = \xi + x_0, \quad y = \eta + y_0.$$

Note 2. The given method cannot be applied if the lines $a_1x + b_1y + c_1 = 0$ and $a_2x + b_2y + c_2 = 0$ are parallel. But in this case

$$\frac{a_2}{a_1} = \frac{b_2}{b_1} = \lambda,$$

and equation (2) can be written in the form

$$\frac{dy}{dx} = f\left[\frac{a_1x + b_1y + c_1}{\lambda(a_1x + b_1y) + c_2}\right] = F(a_1x + b_1y), \tag{3}$$

and an equation of the form (3) has already been considered in §1.

Sometimes an equation can be reduced to a homogeneous one by a change of variable $y = z^\alpha$. This happens when all the terms in the equation have the same degree, if the variable x is given degree 1, the variable y the degree α and the derivative dy/dx the degree $\alpha - 1$.

Example. Solve the equation

$$xy' = \sqrt{x^2 - y^2} + y.$$

Solution. We write the equation in the form

$$y' = \sqrt{1 - \left(\frac{y}{x}\right)^2} + \frac{y}{x}.$$

Since the equation is homogeneous, we put $u = y/x$ or $y = ux$. Then $y' = xu' + u$. Substituting into the equation the expressions for y and y', we obtain

$$x\frac{du}{dx} = \sqrt{1 - u^2}.$$

We separate the variables:

$$\frac{du}{\sqrt{1 - u^2}} = \frac{dx}{x}.$$

By integrating we find that

$$\text{arc sin } u = \ln |x| + \ln C_1 \,(C_1 > 0) \quad \text{or} \quad \text{arc sin } u = \ln C_1|x|.$$

Since $C_1|x| = \pm C_1 x$, by putting $\pm C_1 = C$, we obtain

$$\text{arc sin } u = \ln Cx,$$

where $|\ln Cx| \leqslant \frac{1}{2}\pi$ or $e^{-\frac{1}{2}\pi} \leqslant Cx \leqslant e^{\frac{1}{2}\pi}$. Replacing u by y/x, we have the general integral:

$$\text{arc sin } \frac{y}{x} = \ln Cx.$$

So the general solution is:

$$y = x \sin \ln Cx.$$

In separating the variables we divided both sides of the equation by the product $x\sqrt{(1 - u^2)}$, so we may have lost the solutions which make either of its factors zero. Let us now put $x = 0$ and $\sqrt{(1 - u^2)} = 0$. But $x = 0$ is not a solution of the equation, and from the second we obtain $1 - \dfrac{y^2}{x^2} = 0$, so $y = \pm x$. We may verify immediately that the functions $y = -x$ and $y = x$ are solutions of the equation. The functions $y = -x$ and $y = x$ are singular solutions of the given equation.

Example. Solve the equation

$$(x + y - 2)\, dx + (x - y + 4)\, dy = 0. \tag{1}$$

Solution. We consider the system of linear algebraic equations:

$$\left.\begin{aligned} x + y - 2 = 0 \\ x - y + 4 = 0 \end{aligned}\right\}.$$

The determinant of this system is

$$\Delta = \begin{vmatrix} 1 & 1 \\ 1 & -1 \end{vmatrix} = -2 \neq 0.$$

The system has a unique solution $x_0 = -1$, $y_0 = 3$. We put $x = \xi - 1$, $y = \eta + 3$. Then equation (1) takes the form

$$(\xi + \eta)\, d\xi + (\xi - \eta)\, d\eta = 0. \tag{2}$$

Equation (2) is homogeneous. Putting $\eta = u\xi$, we obtain

$$(\xi + \xi u)\, d\xi + (\xi - \xi u)(\xi\, du + u\, d\xi) = 0.$$

Hence

$$(1 + 2u - u^2)\, d\xi + \xi(1 - u)\, du = 0.$$

We separate the variables:

$$\frac{d\xi}{\xi} + \frac{1 - u}{1 + 2u - u^2}\, du = 0.$$

Integrating, we find that

$$\ln|\xi| + \tfrac{1}{2}\ln|1 + 2u - u^2| = \ln C; \quad \xi^2(1 + 2u - u^2) = C.$$

Returning to the variables x, y:

$$(x + 1)^2\left[1 + 2\frac{y - 3}{x + 1} - \frac{(y - 3)^2}{(x + 1)^2}\right] = C$$

or

$$x^2 + 2xy - y^2 - 4x + 8y = C.$$

Example. Solve the equation

$$(x + y + 1)\, dx + (2x + 2y - 1)\, dy = 0.$$

Solution. The system of linear algebraic equations

$$\left.\begin{aligned} x + y + 1 = 0 \\ 2x + 2y - 1 = 0 \end{aligned}\right\}$$

is inconsistent. The determinant of the system is

$$\Delta = \begin{vmatrix} 1 & 1 \\ 2 & 2 \end{vmatrix} = 0.$$

In this case the method used in the preceding example will not work. To integrate the equation we apply the substitution

$$x + y = z, \quad dy = dz - dx.$$

The equation now takes the form

$$(2 - z)\, dx + (2z - 1)\, dz = 0.$$

Separating the variables, we obtain

$$dx - \frac{2z - 1}{z - 2}\, dz = 0.$$

Hence

$$x - 2z - 3 \ln |z - 2| = C.$$

Returning to the variables x, y, we obtain the general integral of the given equation: $x + 2y + 3 \ln |x + y - 2| = C$.

Example. Solve the equation

$$(x^2y^2 - 1)\, dy + 2xy^3\, dx = 0.$$

Solution. We make the substitution $y = z^\alpha$, $dy = \alpha z^{\alpha-1}\, dz$, where α is a number which we shall choose later. Substituting into the equation the expressions for y and dy, we obtain

$$(x^2z^{2\alpha} - 1)\alpha z^{\alpha-1}\, dz + 2xz^{3\alpha}\, dx = 0$$

or

$$(x^2z^{3\alpha-1} - z^{\alpha-1})\alpha\, dz + 2xz^{3\alpha}\, dx = 0.$$

We observe that $x^2z^{3\alpha-1}$ has degree $2 + 3\alpha - 1 = 3\alpha + 1$, $z^{\alpha-1}$ has degree $\alpha - 1$, $xz^{3\alpha}$ has degree $1 + 3\alpha$. The new equation is homogeneous if all the terms have the same degree, i.e. if $3\alpha + 1 = \alpha - 1$. Hence $\alpha = -1$.

So we have $y = 1/z$, and the original equation takes the form

$$\left(\frac{1}{z^2} - \frac{x^2}{z^4}\right) dz + 2\frac{x}{z^3}\, dx = 0$$

or

$$(z^2 - x^2)\, dz + 2zx\, dx = 0,$$

We now put $z = ux$, $dz = u\,dx + x\,du$. Then this equation takes the form $(u^2 - 1)(u\,dx + x\,du) + 2u\,dx = 0$. Hence

$$u(u^2 + 1)\,dx + x(u^2 - 1)\,du = 0.$$

Separating the variables in this equation, we obtain

$$\frac{dx}{x} + \frac{u^2 - 1}{u^3 + u}\,du = 0.$$

Integrating, we find that

$$\ln|x| + \ln(u^2 + 1) - \ln|u| = \ln C$$

or

$$\frac{x(u^2 + 1)}{u} = C.$$

Replacing u by $1/xy$, we obtain the general integral of the given equation: $1 + x^2y^2 = Cy$.

The equation has the obvious solution $y = 0$, which is obtained from the general integral as $C \to \infty$, if it is written in the form $y = (1 + x^2y^2)/C$, and then we proceed to the limit as $C \to \infty$. Thus the function $y = 0$ is a particular integral of the original equation.

Integrate the equations:

131. $4x - 3y + y'(2y - 3x) = 0.$

132. $xy' = y + \sqrt{y^2 - x^2}.$

133. $4x^2 - xy + y^2 + y'(x^2 - xy + 4y^2) = 0.$

134. $4x^2 + xy - 3y^2 + y'(-5x^2 + 2xy + y^2) = 0.$

135. $y' = \dfrac{2xy}{3x^2 - y^2}.$

136. $2xy'(x^2 + y^2) = y(y^2 + 2x^2).$

137. $xy' = \sqrt{y^2 - x^2}.$

138. $ax^2 + 2bxy + cy^2 + y'(bx^2 + 2cxy + fy^2) = 0.$

139. $(y^4 - 3x^2)\,dy = -xy\,dx.$

140. $y^3\,dx + 2(x^2 - xy^2)\,dy = 0.$

141. $(y - xy')^2 = x^2 + y^2.$

142. $3x + y - 2 + y'(x - 1) = 0.$

143. $2x + 2y - 1 + y'(x + y - 2) = 0.$

144. $(3y - 7x + 7)\,dx - (3x - 7y - 3)\,dy = 0.$

145. $(y + y\sqrt{x^2y^4 + 1})\,dx + 2x\,dy = 0.$

146. $4xy^2\,dx + (3x^2y - 1)\,dy = 0.$

147. $(x + y^3)\,dx + (3y^5 - 3y^2x)\,dy = 0.$

148. $2(x^2y + \sqrt{1 + x^4y^2})\,dx + x^3\,dy = 0.$

149. $y\cos x\,dx + (2y - \sin x)\,dy = 0.$

150. $\left(x - y\cos\frac{y}{x}\right)dx + x\cos\frac{y}{x}\,dy = 0.$

151. $y^3\,dy + 3y^2x\,dx + 2x^3\,dx = 0.$

152. $y\,dx + (2\sqrt{xy} - x)\,dy = 0.$

153. Find a curve having the property that the length of the perpendicular dropped from the origin to a tangent is equal to the abscissa of the point of contact.

154. Determine a curve for which the ratio of the intercept cut by the tangent on OY to the radius vector is constant.

155. Using rectangular coordinates, find the form of a mirror which reflects all rays coming from a given point in a given direction.

156. Find a curve for which the segment cut by the normal at any point of the curve on the axis OY is equal to the distance of the point from the origin.

157. Find a curve for which the product of the abscissa of any point and the segment cut by the normal on the axis OX is equal to twice the square of the distance of this point from the origin.

§6. LINEAR EQUATIONS OF THE FIRST ORDER BERNOULLI EQUATIONS

An equation is called a *linear differential equation of the first order* if it is linear in the unknown function and its derivative. It has the form

$$\frac{dy}{dx} + p(x)y = q(x), \qquad (1)$$

when $p(x)$ and $q(x)$ are continuous functions of x in the range in which equation (1) is to be integrated.

If $q(x) \equiv 0$, equation (1) is called homogeneous linear. It is an equation with separable variables and has the general solution

$$y = ce^{-\int p(x)\,dx}. \tag{2}$$

The general solution of a non-homogeneous equation can be found by the method of variation of parameters, i.e. we look for a solution of (1) in the form

$$y = c(x)e^{-\int p(x)\,dx},$$

where $c(x)$ is a new unknown function of x.

Equation (1) can also be integrated in the following way. We put

$$y = u(x)v(x). \tag{3}$$

Substituting (3) in (1), we obtain

$$u'v + u(pv + v') = q(x). \tag{4}$$

Defining $v(x)$ by the condition $v' + pv = 0$, we obtain from (4) a function $u(x)$, and consequently a solution of equation (1), $y = uv$. For $v(x)$ any particular solution of the equation $v' + pv = 0$ can be chosen.

Note. It can happen that a differential equation is linear for x as a function of y. The normal form of such an equation is

$$\frac{dx}{dy} + r(y)x = \phi(y).$$

Example. Solve the equation

$$y' + 2xy = 2xe^{-x^2}. \tag{5}$$

Solution. We use the method of variation. We consider the homogeneous equation corresponding to the given non-homogeneous equation

$$y' + 2xy = 0.$$

This is an equation with separable variables. Its general solution has the form

$$y = ce^{-x^2}.$$

We look for the general solution of the non-homogeneous equation in the form

$$y = c(x)e^{x^2-}, \tag{6}$$

where $c(x)$ is an unknown function of x. Substituting (6) into (5), we obtain $c'(x) = 2x$. Hence $c(x) = x^2 + c$. So the general solution of the non-homogeneous equation is

$$y = (x^2 + c)e^{-x^2}.$$

Example. Solve the equation

$$\frac{dy}{dx} = \frac{1}{x \cos y + \sin 2y}.$$

Solution. The given equation is linear, if x is regarded as a function of y:

$$\frac{dx}{dy} - x \cos y = \sin 2y. \tag{7}$$

We look for the general solution of the given equation in the form $x = u(y)v(y)$. We have

$$\frac{dx}{dy} = v\frac{du}{dy} + u\frac{dv}{dy}.$$

Substituting x and dx/dy into equation (7), we find that

$$v\frac{du}{dy} + u\left(\frac{dv}{dy} - v \cos y\right) = \sin 2y.$$

We find the function $v(y)$ from the condition

$$\frac{dv}{dy} - v \cos y = 0.$$

We choose any particular solution of this equation, for example $v(y) = e^{\sin y}$. Then

$$e^{\sin y}\frac{du}{dy} = \sin 2y.$$

Hence

$$u = \int e^{-\sin y} \sin 2y\, dy = -2e^{-\sin y}(1 + \sin y) + c.$$

Consequently the general solution is

$$x = ce^{\sin y} - 2 \sin y - 2.$$

A Bernoulli equation has the form

$$\frac{dy}{dx} + p(x)y = q(x)y^n, \tag{8}$$

where $n \neq 0, 1$, since for $n = 0$ and $n = 1$ this equation is linear. By a change of variables $z = 1/y^{n-1}$ equation (8) reduces to a linear equation. However, in practice it is more convenient to solve a Bernoulli equation by means of a substitution $y = u(x)v(x)$ (not reducing it to a linear equation).

Example. Solve the Bernoulli equation

$$xy' + y = y^2 \ln x.$$

Solution. We put $y = u(x)v(x)$. Then we have

$$xvu' + u(xv' + v) = u^2v^2 \ln x.$$

We find the function $v(x)$ as a particular solution of the equation $xv' + v = 0$. We have $v(x) = 1/x$. Then $u' = (u^2/x^2) \ln x$. Hence, separating the variables and integrating, we obtain

$$-\frac{1}{u} = \int \frac{\ln x}{x^2}\, dx = -\frac{\ln x}{x} - \frac{1}{x} - c,$$

$$u = \frac{x}{1 + cx + \ln x}.$$

So the general solution of the equation has the form

$$y = \frac{1}{1 + cx + \ln x}.$$

Integrate the equations

158. $y' + 2y = x^2 + 2x.$

159. $(x^2 + 2x - 1)y' - (x + 1)y = x - 1.$

160. $y'x \ln x - y = x^3(3 \ln x - 1).$

161. $(a^2 - x^2)y' + xy = a^2.$

162. $2xy' - y = 3x^2.$

163. $(x + 1)\, dy - [2y + (x + 1)^4]\, dx = 0.$

164. $y' = \dfrac{1}{x \sin y + 2 \sin 2y}.$

165. $y' - 2xy = 2xe^{x^2}.$

166. $x(x^3 + 1)y' + (2x^3 - 1)y = \dfrac{x^3 - 2}{x}.$

167. $y' + y \cos x = \sin x \cos x,\ y|_{x=0} = 1.$

168. $y'x \ln x - y(1 + \ln x) + \frac{1}{2}\sqrt{x}(2 + \ln x) = 0.$

169. $3xy' - 2y = \dfrac{x^3}{y^2}.$

170. $8xy' - y = -\dfrac{1}{y^3\sqrt{x+1}}.$

171. $(xy + x^2y^3)y' = 1.$

172. $x^2y' + 2x^3y = y^2(1 + 2x^2).$

173. $y' = \dfrac{2xy}{x^2 - y^2 - a^2}.$

174. $2y' \sin x + y \cos x = y^3(x \cos x - \sin x).$

175. $y' = \dfrac{3x^2}{x^3 + y + 1}.$

176. $y' + y\dfrac{x + \frac{1}{2}}{x^2 + x + 1} = \dfrac{(1 - x^2)y^2}{(x^2 + x + 1)^{3/2}}.$

177. $3y' + y\dfrac{x^2 + a^2}{x(x^2 - a^2)} = \dfrac{1}{y^2}\dfrac{x(3x^2 - a^2)}{x^2 - a^2}.$

178. $(1 + x^2)y' = xy + x^2y^2.$

179. $y' + \dfrac{y}{x+1} = -\dfrac{1}{2}(x + 1)^3y^3.$

180. $(x^2 + y^2 + 1)\,dy + xy\,dx = 0.$

181. $y' = \dfrac{y}{2y \ln y + y - x}.$

182. $x(x - 1)y' + y = x^2(2x - 1).$

183. $y' - y \tan x = \sec x, \qquad y|_{x=0} = 0.$

184. $y' \cos y + \sin y = x + 1.$

185. $y' + \sin y + x \cos y + x = 0.$

186. $y' - \dfrac{ny}{x+1} = e^x(1 + x)^n.$

187. $\int_0^1 \phi(\alpha x)\,d\alpha = n\phi(x).$

188. $y' + x \sin 2y = xe^{-x^2} \cos^2 y$
(substitute $t = \tan y$).

In the following problems find the solutions of the equations which satisfy the given conditions:

189. $y' - 2xy = \cos x - 2x \sin x,\qquad y$ bounded as $x \to \infty$.

190. $2\sqrt{x}y' - y = -\sin\sqrt{x} - \cos\sqrt{x}$,
y bounded as $x \to +\infty$.

191. $y' - y \ln 2 = 2^{\sin x}(\cos x - 1) \ln 2$,
y bounded as $x \to +\infty$.

192. $2x^2y' - xy = 2x \cos x - 3 \sin x,\qquad y \to 0$ as $x \to +\infty$.

193. $y' \sin x - y \cos x = -\dfrac{\sin^2 x}{x^2},\qquad y \to 0$ as $x \to \infty$.

194. $(1 + x^2) \ln (1 + x^2)y' - 2xy = \ln (1 + x^2) - 2x \arctan x$,

$y \to -\dfrac{\pi}{2}$ as $x \to -\infty$.

195. $y' - e^x y = \dfrac{1}{x^2} \sin\dfrac{1}{x} - e^x \cos\dfrac{1}{x},\qquad y \to 2$ as $x \to -\infty$.

196. $y' - y \ln x = -(1 + 2 \ln x)x^{-x},\qquad y \to 0$ as $x \to +\infty$.

§7. TOTAL DIFFERENTIAL EQUATIONS INTEGRATING FACTORS

A differential equation of the form

$$M(x, y)\, dx + N(x, y)\, dy = 0 \tag{1}$$

is called a total differential equation if its left hand side represents the total differential of some function $u(x, y)$

$$M\, dx + N\, dy \equiv du \equiv \frac{\partial u}{\partial x}\, dx + \frac{\partial u}{\partial y}\, dy.$$

A necessary and sufficient condition that equation (1) is a total differential equation is

$$\frac{\partial M}{\partial y} \equiv \frac{\partial N}{\partial x} \text{ (in some region } D \text{ of the } xy\text{-plane).} \tag{2}$$

The general integral of equation (1) has the form $u(x, y) = C$ or

$$\int_{x_0}^{x} M(x, y)\,dx + \int_{y_0}^{y} N(x_0, y)\,dy = C. \tag{3}$$

Example. Solve the differential equation

$$(\sin xy + xy \cos xy)\,dx + x^2 \cos xy\,dy = 0.$$

Solution. We verify that the given equation is a total differential equation. We have

$$\frac{\partial M}{\partial y} = \frac{\partial}{\partial y}(\sin xy + xy \cos xy) = x \cos xy + x \cos xy - x^2 y \sin xy$$
$$= 2x \cos xy - x^2 y \sin xy,$$

$$\frac{\partial N}{\partial x} = \frac{\partial}{\partial x}(x^2 \cos xy) = 2x \cos xy - x^2 y \sin xy,$$

i.e.

$$\frac{\partial M}{\partial y} \equiv \frac{\partial N}{\partial x}.$$

Condition (2) is satisfied. Further,

$$u(x, y) = \int_{x_0}^{x} (\sin xy + xy \cos xy)\,dx + \int_{y_0}^{y} x_0^2 \cos x_0 y\,dy$$

$$= x \sin xy \Big|_{x_0}^{x} + x_0 \sin x_0 y \Big|_{y_0}^{y} = x \sin xy - x_0 \sin x_0 y_0;$$

so

$$x \sin xy = C + x_0 \sin x_0 y_0 \quad \text{or} \quad x \sin xy = C_1.$$

To integrate certain differential equations it is possible to group the terms so that we obtain easily integrable combinations.

Example. Solve the differential equation

$$(x^3 + xy^2)\,dx + (x^2 y + y^3)\,dy = 0. \tag{2'}$$

Solution. Here $\dfrac{\partial M}{\partial y} = 2xy$, $\dfrac{\partial N}{\partial x} = 2xy$, so condition (2) is satisfied, and consequently the given equation is a total differential equation. This equation is easily reduced to the form $du = 0$ by a regrouping of its terms. To this end, we rewrite it thus:

$$x^3\,dx + xy(y\,dx + x\,dy) + y^3\,dy = 0.$$

Here

$$x^3\,dx = d\left(\frac{x^4}{4}\right),\quad xy(y\,dx + x\,dy) = xy\,d(xy) = d\left(\frac{(xy)^2}{2}\right),$$

$$y^3\,dy = d\left(\frac{y^4}{4}\right).$$

Therefore equation (2′) can be written in the form

$$d\left(\frac{x^4}{4}\right) + d\left(\frac{(xy)^2}{2}\right) + d\left(\frac{y^4}{4}\right) = 0$$

or

$$d\left[\frac{x^4}{4} + \frac{(xy)^2}{2} + \frac{y^4}{4}\right] = 0. \tag{2″}$$

Consequently,

$$x^4 + 2(xy)^2 + y^4 = C$$

is the general integral of equation (2″), and therefore of equation (2′).

In certain cases, where the left hand side of equation (1) is not a total differential, it turns out that we can choose a function $\mu(x, y)$ by which to multiply the left hand side of equation (1) so that it becomes a total differential

$$du = \mu M\,dx + \mu N\,dy. \tag{4}$$

Such a function $\mu(x, y)$ is called an integrating factor. From the definition of an integrating factor, we have

$$\frac{\partial}{\partial y}(\mu M) = \frac{\partial}{\partial x}(\mu N)$$

or

$$N\frac{\partial \mu}{\partial x} - M\frac{\partial \mu}{\partial y} = \left(\frac{\partial M}{\partial y} - \frac{\partial N}{\partial x}\right)\mu,$$

hence

$$N\frac{\partial \ln \mu}{\partial x} - M\frac{\partial \ln \mu}{\partial y} = \frac{\partial M}{\partial y} - \frac{\partial N}{\partial x}. \tag{5}$$

To find an integrating factor, we have obtained a partial differential equation.

We single out certain special cases where it is comparatively simple to find a solution of equation (5), i.e. to find an integrating factor.

1. $\mu = \mu(x)$. Then $\dfrac{\partial \mu}{\partial y} = 0$ and equation (5) takes the form

$$\frac{d \ln \mu}{dx} = \frac{\dfrac{\partial M}{\partial y} - \dfrac{\partial N}{\partial x}}{N}. \tag{6}$$

For the existence of an integrating factor not depending on y, it is necessary and sufficient that the right hand side of (6) should be a function of x only. In this case $\ln \mu$ can be found by quadrature.

Example. Consider the equation

$$(x + y^2)\, dx - 2xy\, dy = 0.$$

Solution. Here

$$M = x + y^2, \quad N = -2xy.$$

We have

$$\frac{\dfrac{\partial M}{\partial y} - \dfrac{\partial N}{\partial x}}{N} = \frac{2y + 2y}{-2xy} = -\frac{2}{x}.$$

Consequently

$$\frac{d \ln \mu}{dx} = -\frac{2}{x}, \quad \ln \mu = -2 \ln |x|, \quad \mu = \frac{1}{x^2}.$$

The equation

$$\frac{x + y^2}{x^2}\, dx - 2\frac{xy}{x^2}\, dy = 0$$

is a total differential equation. Its left hand side can be expressed in the form

$$\frac{dx}{x} - \frac{2xy\, dy - y^2\, dx}{x^2} = 0.$$

Hence

$$d\left(\ln |x| - \frac{y^2}{x}\right) = 0$$

and the general integral of the given equation is

$$x = Ce^{\frac{y^2}{x}}.$$

2. Similarly, if $\left(\frac{\partial N}{\partial x} - \frac{\partial M}{\partial y}\right)\frac{1}{M}$ is a function of y only, equation (1) has an integrating factor $\mu = \mu(y)$ depending only on y.

Example. $2xy \ln y\, dx + (x^2 + y^2\sqrt{y^2+1})dy = 0$.

Solution. Here $M = 2xy \ln y$, $N = x^2 + y^2\sqrt{y^2+1}$. We have

$$\frac{\frac{\partial N}{\partial x} - \frac{\partial M}{\partial y}}{M} = \frac{2x - 2x(\ln y + 1)}{2xy \ln y} = -\frac{1}{y}.$$

Consequently,

$$\frac{d \ln \mu}{dy} = -\frac{1}{y}, \quad \mu = \frac{1}{y}.$$

The equation

$$\frac{2xy \ln y\, dx}{y} + \frac{x^2 + y^2\sqrt{y^2+1}}{y}\,dy = 0$$

is a total differential equation. It can be written in the form $d(x^2 \ln y) + y\sqrt{(y^2+1)}dy = 0$ and the general integral is

$$x^2 \ln y + \frac{1}{3}(y^2+1)^{3/2} = C.$$

Integrate the equations

197. $x(2x^2 + y^2) + y(x^2 + 2y^2)y' = 0$.

198. $(3x^2 + 6xy^2)\,dx + (6x^2y + 4y^3)\,dy = 0$.

199. $\left(\frac{x}{\sqrt{x^2+y^2}} + \frac{1}{x} + \frac{1}{y}\right)dx + \left(\frac{y}{\sqrt{x^2+y^2}} + \frac{1}{y} - \frac{x}{y^2}\right)dy = 0$.

200. $\left(3x^2 \tan y - \frac{2y^3}{x^3}\right)dx + \left(x^3 \sec^2 y + 4y^3 + \frac{3y^2}{x^2}\right)dy = 0$.

201. $\left(2x + \frac{x^2+y^2}{x^2y}\right)dx = \frac{x^2+y^2}{xy^2}\,dy$.

202. $\left(\frac{\sin 2x}{y} + x\right)dx + \left(y - \frac{\sin^2 x}{y^2}\right)dy = 0$.

203. $(3x^2 - 2x - y)\,dx + (2y - x + 3y^2)\,dy = 0$.

204. $\left(\frac{xy}{\sqrt{1+x^2}} + 2xy - \frac{y}{x}\right)dx + (\sqrt{1+x^2} + x^2 - \ln x)\,dy = 0$.

205. $\dfrac{x\,dx + y\,dy}{\sqrt{x^2 + y^2}} + \dfrac{x\,dy - y\,dx}{x^2} = 0.$

206. $\left(\sin y + y \sin x + \dfrac{1}{x}\right) dx + \left(x \cos y - \cos x + \dfrac{1}{y}\right) dy = 0.$

207. $\dfrac{y + \sin x \cos^2 xy}{\cos^2 xy}\, dx + \left(\dfrac{x}{\cos^2 xy} + \sin y\right) dy = 0.$

208. $\dfrac{2x\,dx}{y^3} + \dfrac{(y^2 - 3x^2)\,dy}{y^4} = 0, \quad y|_{x=1} = 1.$

209. $\{n \cos (nx + my) - m \sin (mx + ny)\}\, dx$
$+ \{m \cos (nx + my) - n \sin (mx + ny)\}\, dy = 0.$

210. $\dfrac{x\,dx + y\,dy}{\sqrt{(x^2 + y^2)(1 - x^2 - y^2)}} + \left(\dfrac{1}{y\sqrt{y^2 - x^2}} + \dfrac{e^{\frac{x}{y}}}{y^2}\right)$
$\times (y\,dx - x\,dy) = 0.$

211. $\left(\dfrac{1}{y} \sin \dfrac{x}{y} - \dfrac{y}{x^2} \cos \dfrac{y}{x} + 1\right) dx$
$+ \left(\dfrac{1}{x} \cos \dfrac{y}{x} - \dfrac{x}{y^2} \sin \dfrac{x}{y} + \dfrac{1}{y^2}\right) dy = 0.$

212. $y(x^2 + y^2 + a^2)\, dy + x(x^2 + y^2 - a^2)\, dx = 0.$

213. $(x^2 + y^2 + 1)\, dx - 2xy\, dy = 0, \quad \mu = \phi(y^2 - x^2).$

214. $(1 - x^2 y)\, dx + x^2(y - x)\, dy = 0.$

215. $y(x^2 + y^2)\, dx + x(x\,dy - y\,dx) = 0, \quad \mu = \phi[y(x^2 + y^2)].$

216. $x\,dx + y\,dy + x(x\,dy - y\,dx) = 0, \quad \mu = \phi(x^2 + y^2).$

217. $(x^2 + y)\, dx - x\, dy = 0, \quad \mu = \phi(x).$

218. $(x + y^2)\, dx - 2xy\, dy = 0, \quad \mu = \phi(x).$

219. $(2x^2 y + 2y + 5)\, dx + (2x^3 + 2x)\, dy = 0, \quad \mu = \phi(x).$

220. $(x^4 \ln x - 2xy^3)\, dx + 3x^2 y^2\, dy = 0, \quad \mu = \phi(x).$

221. $(x + \sin x + \sin y)\, dx + \cos y\, dy = 0, \quad \mu = \phi(x).$

222. $(2xy^2 - 3y^3)\, dx + (7 - 3xy^2)\, dy = 0, \quad \mu = \phi(y).$

223. $(3y^2 - x)\, dx + (2y^3 - 6xy)\, dy = 0, \quad \mu = \phi(x + y^2).$

§8. DIFFERENTIAL EQUATIONS OF THE FIRST ORDER NOT SOLVED FOR THE DERIVATIVE

1. Equations of the first order and of degree n in y'.

$$(y')^n + p_1(x, y)(y')^{n-1} + \dots + p_{n-1}(x, y)y' + p_n(x, y) = 0. \quad (1)$$

We solve this equation for y'. Let

$$y' = f_1(x, y), \quad y' = f_2(x, y), \dots, y' = f_k(x, y) \; (k \leqslant n) \quad (2)$$

be the real solutions of equation (1).

The general integral of equation (1) consists of the union of the integrals

$$\Phi_1(x, y, C) = 0, \quad \Phi_2(x, y, C) = 0, \dots, \Phi_k(x, y, C) = 0, \quad (3)$$

where $\Phi_i(x, y, C) = 0$ is an integral of the equation

$$y' = f_i(x, y) \; (i = 1, \dots, k).$$

Thus through each point of the region at which y' takes real values there pass k integral curves,

Example. Solve the equation

$$yy'^2 + (x - y)y' - x = 0.$$

Solution. Solving this equation for y', we have

$$y' = \frac{y - x \pm \sqrt{(x - y)^2 + 4xy}}{2y}$$

or $y' = 1$, $y' = -x/y$, and hence

$$y = x + C, \quad y^2 + x^2 = C^2.$$

Integrate the following equations:

224. $y'^2 - (2x + y)y' + (x^2 + xy) = 0.$

225. $xy'^2 + 2xy' - y = 0.$

226. $4y'^2 - 9x = 0.$

227. $y'^2 - 2yy' = y^2(e^x - 1).$

228. $x^2y'^2 + 3xyy' + 2y^2 = 0.$

229. $xy'^2 - 2yy' + x = 0.$

230. $(x^2 - 2xy)y'^2 - 2xyy' + (y^2 - 2xy) = 0.$

231. $y'^3 + (x + 2)e^y = 0.$

232. $y'^3 - yy'^2 - x^2y' + x^2y = 0.$

2. Equations of the form $f(y, y') = 0$ or $f(x, y') = 0$.

If these equations are easily soluble for y', then, solving them, we obtain equations with separable variables. Certain other cases present greater interest.

(*a*) An equation of the form $f(y, y') = 0$ easily soluble for y:

$$y = \phi(y').$$

We put $y' = p$. Then $y = \phi(p)$. Differentiating this equation and replacing dy by $p\,dx$, we obtain

$$p\,dx = \phi'(p)\,dp,$$

so

$$dx = \frac{\phi'(p)}{p}\,dp \quad \text{and} \quad x = \int \frac{\phi'(p)}{p}\,dp + C.$$

We obtain the general solution of the equation in the parametric form

$$\left.\begin{aligned} x &= \int \frac{\phi'(p)}{p}\,dp + C \\ y &= \phi(p). \end{aligned}\right\}.$$

Example. $y = a\left(\frac{dy}{dx}\right)^2 + b\left(\frac{dy}{dx}\right)^3$ $(a, b$ – constants).

Solution. $dy/dx = p$. Then $y = ap^2 + bp^3$,

$$dy = 2ap\,dp + 3bp^2\,dp \quad \text{or} \quad p\,dx = 2ap\,dp + 3bp^2\,dp.$$

Hence

$$dx = 2a\,dp + 3bp\,dp \quad \text{and} \quad x = 2ap + \tfrac{3}{2}bp^2 + C.$$

The general solution is

$$\left.\begin{aligned} x &= 2ap + \tfrac{3}{2}bp^2 + C \\ y &= ap^2 + bp^3 \end{aligned}\right\}.$$

(*b*) An equation $f(y, y') = 0$ not soluble (or soluble with difficulty) for y or y', but admitting parametric expressions for y and y' in terms of some parameter t:

$$y = \phi(t), \quad p = \psi(t) \left(p = \frac{dy}{dx}\right).$$

Then $dy = p\,dx = \psi(t)\,dx$.

On the other hand $dy = \phi'(t)\,dt$, so that $\psi(t)\,dx = \phi'(t)\,dt$ and $dx = \dfrac{\phi'(t)}{\psi(t)}\,dt$. Hence

$$x = \int \frac{\phi'(t)}{\psi(t)}\,dt.$$

Thus we obtain the general solution of the given differential equation in the parametric form

$$\left.\begin{aligned} x &= \int \frac{\phi'(t)}{\psi(t)}\,dt + C \\ y &= \phi(t) \end{aligned}\right\}.$$

Example. $y^{2/3} + (y')^{2/3} = 1.$

Solution. We put

$$y = \cos^3 t, \quad y' = p = \sin^3 t,$$

$$dx = \frac{dy}{p} = \frac{-3\cos^2 t \sin t\,dt}{\sin^3 t} = -3\,\frac{\cos^2 t}{\sin^2 t}\,dt.$$

Hence

$$x = \int \left(3 - \frac{3}{\sin^2 t}\right) dt = 3t + 3\cot t + C.$$

The general solution is

$$\left.\begin{aligned} x &= 3t + 3\cot t + C \\ y &= \cos^3 t \end{aligned}\right\}.$$

(*c*) An equation $f(x, y') = 0$. Suppose this equation is soluble for x: $x = \phi(y')$. Putting $y' \equiv p$, we obtain

$$dx = \phi'(p)\,dp.$$

But $dx = \dfrac{dy}{p}$ and consequently $\dfrac{dy}{p} = \phi'(p)\,dp$, so $dy = p\phi'(p)\,dp$ and $y = \int p\phi'(p)\,dp + C$. Thus we find that

$$\left.\begin{aligned} x &= \phi(p) \\ y &= \int \phi'(p)p\,dp + C \end{aligned}\right\}$$ is the general solution of the equation in parametric form (p is the parameter).

Example. $a\dfrac{dy}{dx} + b\left(\dfrac{dy}{dx}\right)^2 = x.$

Solution.

$$\frac{dy}{dx} = p, \quad x = ap + bp^2, \quad dx = a\,dp + 2bp\,dp,$$

$$dy = p\,dx = ap\,dp + 2bp^2\,dp,$$

$$y = \frac{a}{2}p^2 + \frac{2}{3}bp^3 + C.$$

$$\left.\begin{aligned} x &= ap + bp^2 \\ y &= \frac{ap^2}{2} + \frac{2}{3}bp^3 + C \end{aligned}\right\} \text{ is the general solution.}$$

Just as in case (*b*) we may try to solve an equation $f(x, y') = 0$ by introducing a parameter t.

Integrate the following equations:

233. $y = y'^2 e^{y'}$.

234. $y' = e^{\frac{xy'}{y}}$.

235. $x = \ln y' + \sin y'$.

236. $x = y'^2 - 2y' + 2$.

237. $y = y' \ln y'$.

238. $y = \arcsin y' + \ln(1 + y'^2)$.

239. $y = (y' - 1)e^{y'}$.

240. $y'^2 x = e^{\frac{1}{y'}}$.

241. $x(1 + y'^2) = 1$.

242. $x(1 + y'^2)^{3/2} = a$.

243. $y^{2/5} + y'^{2/5} = a^{2/5}$.

244. $y^4 - y'^4 - yy'^2 = 0$.

245. $x = y' + \sin y'$.

246. $y = y'(1 + y' \cos y')$.

3. Equations of Lagrange and Clairaut.

(*a*) A *Lagrange equation* has the form

$$y = x\phi(y') + \psi(y').$$

Putting $y' = p$, differentiating, and replacing dy by $p\,dx$, we reduce this equation to a linear equation for x as a function of p. By finding a solution $x = r(p, C)$ of this equation, we obtain the general solution of the original equation in the parametric form:

$$\left.\begin{aligned} x &= r(p, C) \\ y &= r(p, C)\phi(p) + \psi(p) \end{aligned}\right\} \text{(where } p \text{ is a parameter).}$$

Apart from this, a Lagrange equation can have a singular solution of the form $y = \phi(c)x + \psi(c)$, where c is a root of the equation $c = \phi(c)$.

Example. Integrate the equation

$$y = 2xy' + \ln y'.$$

Solution. We put $y' = p$. Then $y = 2xp + \ln p$. Differentiating, we find that

$$p\,dx = 2p\,dx + 2x\,dp + \frac{dp}{p},$$

hence

$$p\frac{dx}{dp} = -2x - \frac{1}{p} \quad \text{or} \quad \frac{dx}{dp} = -\frac{2}{p}x - \frac{1}{p^2}.$$

We have obtained an equation of the first order, linear with respect to x. Solving it, we find that

$$x = \frac{C}{p^2} - \frac{1}{p}.$$

Substituting this value for x into the expression for y, we obtain finally

$$\left.\begin{aligned} x &= \frac{C}{p^2} - \frac{1}{p} \\ y &= \ln p + \frac{2C}{p} - 2 \end{aligned}\right\}.$$

(*b*) A *Clairaut equation* has the form

$$y = xy' + \psi(y').$$

The method of solution is the same as that for a Lagrange equation. The general solution of a Clairaut equation has the form

$$y = Cx + \psi(C).$$

A Clairaut equation can also have a singular solution, which is obtained by eliminating p from the equations $y = xp + \psi(p)$, $x + \psi'(p) = 0$.

Example. Integrate the equation

$$y = xy' + \frac{a}{2y'} \ (a = \text{const.})$$

Solution. Putting $y' = p$, we obtain

$$y = xp + \frac{a}{2p}.$$

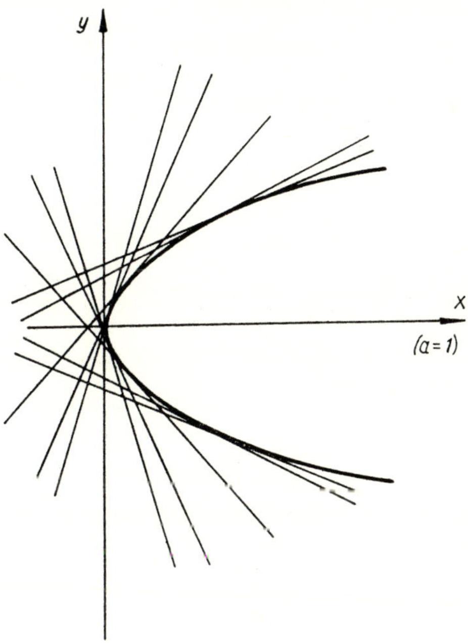

FIGURE 9

Differentiating this equation and replacing dy by $p\,dx$, we find that $p\,dx = p\,dx + x\,dp - \frac{a}{2p^2}\,dp$, and hence $dp\left(x - \frac{a}{2p^2}\right) = 0$.

We investigate both factors of the left hand side of this equation. Equating the first factor to zero, we obtain

$$dp = 0,$$

hence $p = C$ and the general solution of the original equation is

$$y = Cx + \frac{a}{2C}.$$

This is a 1-parameter family of lines.

Equating the second factor to zero, we have

$$x = \frac{a}{2p^2}.$$

Eliminating p from this equation and the equation $y = xp + \frac{a}{2p}$, we obtain $y^2 = 2ax$: this is also a solution of our equation (a singular solution).

From the geometrical point of view, the curve $y^2 = 2ax$ is the envelope of the family of lines given by the general solution (Fig. 9).

Integrate the following equations:

247. $2y = xy' + y' \ln y'$.

248. $y = 2xy' + \ln y'$.

249. $y = x(1 + y') + y'^2$.

250. $y = 2xy' + \sin y'$.

251. $y = xy'^2 - \frac{1}{y'}$.

252. $y = \frac{3}{2}xy' + e^{y'}$.

253. $y = xy' + \frac{a}{y'^2}$.

254. $y = xy' + y'^2$.

255. $xy'^2 - yy' - y' + 1 = 0$.

256. $y = xy' + a\sqrt{1 + y'^2}$.

257. $y = xy' + \frac{ay'}{\sqrt{1 + y'^2}}$.

258. $x = \frac{y}{y'} + \frac{1}{y'^2}$.

259. Find the curve such that any tangent to it forms with the coordinate axes a triangle of constant area $S = 2a^2$.

260. Find the curve such that the segment of any tangent cut off by the coordinate axes has constant length a.

§9. FORMATION OF DIFFERENTIAL EQUATIONS OF FAMILIES OF CURVES. PROBLEMS ON TRAJECTORIES

1. Formation of differential equations of families of curves

Suppose a relation

$$y = \phi(x, a), \tag{1}$$

is given, where a is a parameter. This is the equation of a family of curves. Differentiating (1) with respect to x, we obtain

$$y' = \phi'_x(x, a). \tag{2}$$

If the right hand side of (2) does not contain a, then (2) is the differential equation of the family. Obviously, in this case the relation (1) has the form $y = \phi(x) + a$.

Suppose now that the right hand side of (2) contains a. Solving (1), if possible, for a, we determine a as a function of x and y:

$$a = \psi(x, y). \tag{3}$$

Substituting (3) into (2), we obtain a differential equation of the first order

$$y' = \phi'_x[x, \psi(x, y)]. \tag{4}$$

It is easy to verify that (1) represents the general solution of equation (4).

Let a 1-parameter family of curves be defined by the equation

$$\Phi(x, y, a) = 0, \tag{1'}$$

where a is a parameter. Eliminating the parameter a from the equations

$$\left.\begin{aligned} &\Phi(x, y, a) = 0 \\ &\frac{\partial \Phi}{\partial x} + \frac{\partial \Phi}{\partial y} y' = 0 \end{aligned}\right\},$$

we obtain the differential equation

$$F(x, y, y') = 0$$

of the given family (1').

Suppose now that we have a relation

$$\Phi(x, y, a_1, a_2, \ldots, a_n) = 0, \tag{5}$$

where $a_1, a_2, \ldots, a_n$ are parameters. Differentiating (5) n times with respect to x and eliminating the parameters $a_1, a_2, \ldots, a_n$ from (5) and the equations thus obtained, we arrive at a relation of the form

$$F(x, y, y', y'', \ldots, y^{(n)}) = 0. \tag{6}$$

This is the differential equation of the given n-parameter family of curves (5), in the sense that (5) is the general integral of equation (6).

2. Problems on trajectories

Suppose that a family of plane curves

$$\Phi(x, y, a) = 0, \tag{1}$$

is given, depending on one parameter a.

A curve which at each point makes a constant angle α with the curve of the family (1) which passes through that point is called an isogonal trajectory of the family. If, in particular, $\alpha = \frac{1}{2}\pi$, we have an orthogonal trajectory.

Regarding the family (1) as given, we shall look for its isogonal trajectories.

(*a*) *Orthogonal trajectories*

We form the differential equation of the given family of curves (see section 1). Let this be the equation

$$F(x, y, y') = 0. \tag{2}$$

We can show that the differential equation of the orthogonal trajectories has the form:

$$F\left(x, y, -\frac{1}{y'}\right) = 0. \tag{3}$$

The general integral of this equation

$$\Phi_1(x, y, C) = 0 \tag{4}$$

gives the family of orthogonal trajectories.

Suppose a family of plane curves is given by an equation in polar coordinates

$$\Phi(\rho, \phi, a) = 0, \tag{5}$$

where a is a parameter. Eliminating the parameter a from (5) and $\dfrac{d\Phi}{d\phi} = 0$, we obtain the differential equation of the family (5):

$F(\rho, \phi, \rho') = 0$. Replacing ρ' by $-\rho^2/\rho'$, we obtain the differential equation of the family of orthogonal trajectories

$$F\left(\rho, \phi, -\frac{\rho^2}{\rho'}\right) = 0$$

(see E. L. Ince, *Integration of Ordinary Differential Equations*, p. 32).

(*b*) *Isogonal trajectories*

Let a trajectory cut a curve of the given family at an angle α, where $\tan \alpha = k$. We can show that the differential equation of the isogonal trajectories has the form

$$F\left(x, y, \frac{y' - k}{1 + ky'}\right) = 0.$$

Example. Find the differential equation of the family of hyperbolas

$$\frac{x^2}{a^2} - \frac{y^2}{1^2} = 1.$$

Solution. Differentiating this equation with respect to x, we obtain

$$\frac{2x}{a^2} - 2yy' = 0 \quad \text{or} \quad \frac{x}{a^2} = yy'.$$

Multiplying both sides by x, we have

$$\frac{x^2}{a^2} = xyy'.$$

Substituting into the equation, we find that

$$xyy' - y^2 = 1.$$

This is the differential equation of the family of hyperbolas.

Example. Find the differential equation of the family of curves (a is a parameter)

$$y = a\left(1 - e^{-\frac{x}{a}}\right).$$

Solution. Differentiating both sides of the equation with respect to x, we have

$$y' = e^{-\frac{x}{a}}.$$

From the expression for y' we find that

$$a = -\frac{x}{\ln y'}.$$

Substituting this expression for a into the equation of the family of curves, we obtain

$$y = -\frac{x}{\ln y'}(1 - y')$$

or

$$y \ln y' + x(1 - y') = 0.$$

This is the required differential equation of the given family of curves.

Example. Find the orthogonal trajectories of the family of lines $y = kx$.

Solution. The family $y = kx$ consists of lines passing through the origin. To find the differential equation of the given family we

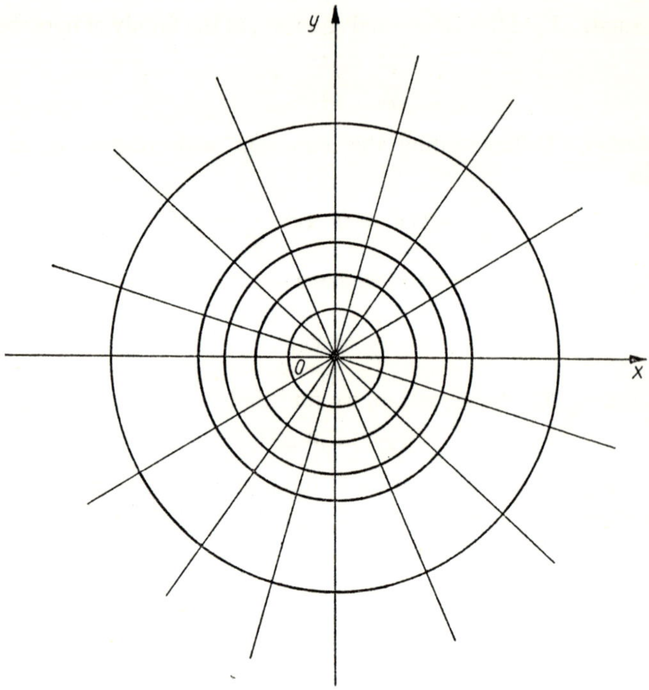

FIGURE 10

differentiate both sides of the equation $y = kx$ with respect to x. We have $y' = k$. Eliminating the parameter k from the system of equations

$$\left.\begin{aligned} y &= kx \\ y' &= k \end{aligned}\right\}$$

we have the differential equation of the family: $xy' = y$. Replacing y' by $-1/y'$, we obtain the differential equation of the orthogonal trajectories: $-x/y' = y$ or $yy' + x = 0$. This is an equation with separable variables; integrating it, we find the equation of the orthogonal trajectories: $x^2 + y^2 = C\,(C \geqslant 0)$. The orthogonal trajectories are circles with their centres at the origin (Fig. 10).

Example. Find the equation of the family of curves orthogonal to the family $x^2 + y^2 = 2ax$.

Solution. The given family of curves consists of circles whose centres lie on the axis OX and which touch the axis OY.

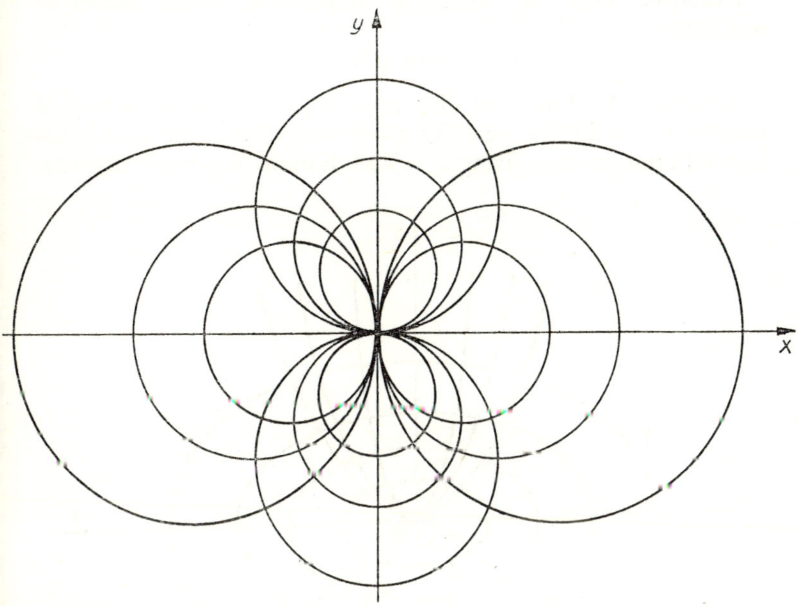

FIGURE 11

Differentiating both sides of the equation of the given family with respect to x, we find that

$$x + yy' = a.$$

Eliminating the parameter a from the equations

$$\left.\begin{aligned} x^2 + y^2 &= 2ax \\ x + yy' &= a \end{aligned}\right\},$$

we obtain the differential equation of the given family:

$$x^2 - y^2 + 2xyy' = 0.$$

The differential equation of the orthogonal trajectories is

$$x^2 - y^2 + 2xy\left(-\frac{1}{y'}\right) = 0 \quad \text{or} \quad y' = \frac{2xy}{x^2 - y^2}.$$

This equation is homogeneous. Integrating it, we find that $x^2 + y^2 = Cy$. The integral curves are circles whose centres lie on the axis OY and which touch the axis OX (Fig. 11).

Example. Find the orthogonal trajectories of the family of parabolas

$$y = ax^2.$$

Solution. We form the differential equation of the family of parabolas. To do this, we differentiate both sides of the given equation with respect to x, giving $y' = 2ax$. Eliminating the parameter a,

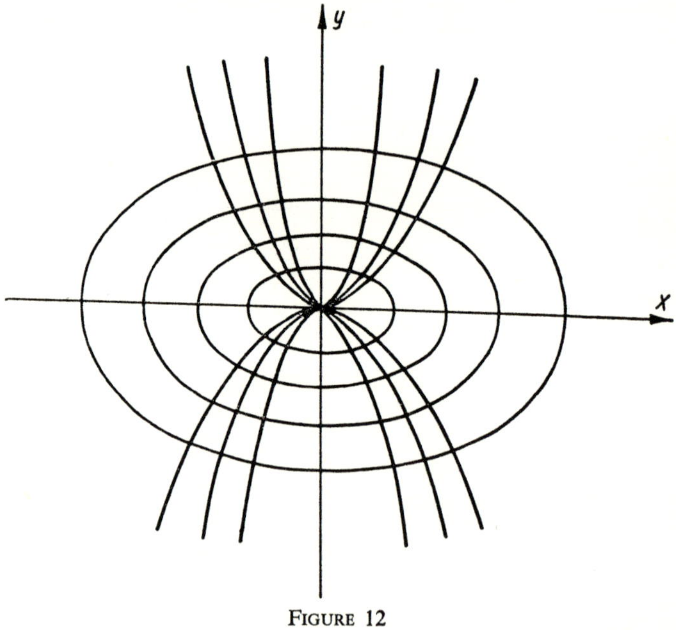

FIGURE 12

we find that $\frac{y'}{y} = \frac{2}{x}$ or $y' = \frac{2y}{x}$ is the differential equation of the given family. Replacing y' by $-1/y'$, we obtain the differential equation of the orthogonal trajectories: $-\frac{1}{y'} = \frac{2y}{x}$. Hence $\frac{dy}{dx} = -\frac{x}{2y}$. Separating the variables: $2y\,dy = -x\,dx$. Integrating, we find that $y^2 = -\frac{1}{2}x^2 + C$ or $\frac{1}{2}x^2 + y^2 = C$, where $C > 0$. So the orthogonal family is a family of ellipses (Fig. 12).

Example. Find the orthogonal trajectories of the family of leminiscates

$$\rho^2 = a \cos 2\phi.$$

Solution. We have

$$\rho^2 = a \cos 2\phi,$$

$$\rho\rho' = -a \sin 2\phi.$$

Eliminating the parameter a, we obtain the differential equation of the given family of curves

$$\rho' = -\rho \tan 2\phi.$$

Replacing ρ' by $-\rho^2/\rho'$, we find the differential equation of the family of orthogonal trajectories:

$$-\frac{\rho^2}{\rho'} = -\rho \tan 2\phi \quad \text{or} \quad \rho' = \rho \cot 2\phi.$$

Hence $\frac{d\rho}{\rho} = \cot 2\phi\, d\phi$. Integrating, we find the equation of the orthogonal trajectories:

$$\rho^2 = C \sin 2\phi$$

The orthogonal trajectories are a family of leminiscates whose axes of symmetry make angles of $\pm 45°$ with the polar axis (Fig. 13).

Form the differential equations of the following families of curves:

261. $y = \frac{a}{x}$.

262. $x^2 - y^2 = ax$.

263. $y = ae^{\frac{x}{a}}$.

264. $y = Cx - C - C^2$.

265. $y = e^x(ax + b)$.

266. $y^2 = 2Cx + C^2$.

267. $y = ax^2 + bx + C$.

268. $y = C_1x + \dfrac{C_2}{x} + C_3.$

269. $(x - a)^2 + (y - b)^2 = 1.$

270. $y = C_1e^x + C_2e^{-x}.$

271. $y = a \sin (x + \alpha).$

Find the orthogonal trajectories of the given families of curves:

272. $y^2 + 2ax = a^2, \quad a > 0.$

273. $\dfrac{x^2}{a^2} + \dfrac{y^2}{a^2 - c^2} = 1,$

$a > c.$

274. $y = ae^{\sigma x}, \quad \sigma = \text{const}.$

275. $\cos y = ae^{-x}.$

276. $x^2 + \frac{1}{2}y^2 = a^2.$

277. $x^2 - y^2 = a^2.$

278. $x^k + y^k = a^k.$

279. $x^2 + y^2 = 2ay.$

280. $x^2 - \frac{1}{3}y^2 = a^2.$

281. $\rho = a(1 + \cos \phi).$

282. $y^2 = 4(x - a).$

FIGURE 13

§10. VARIOUS PROBLEMS

Integrate the following equations:

283. $(y - y^3)\,dx + (2xy^2 - x - ay^2)\,dy = 0.$

284. $y' = (x - y)^2 + 1.$

285. $y'x \sin x + y(\sin x - x \cos x) = \sin x \cos x - x.$

286. $\dfrac{dy}{dx} + y \cos x = y^n \sin 2x.$

287. $(x^3 - 3xy^2)\,dx + (y^3 - 3x^2y)\,dy = 0.$

288. $(5xy - 4y^2 - 6x^2)\,dx + (y^2 - 2xy + 6x^2)\,dy = 0.$

289. $(3xy^2 - x^2)\,dx + (3x^2y - 6y^2 - 1)\,dy = 0.$

290. $(y - xy^2 \ln x)\,dx + x\,dy = 0, \quad \mu = \phi(x\,.\,y).$

291. $(2xye^{x^2} - x \sin x)\,dx + e^{x^2}\,dy = 0.$

292. $2y' + y^2 + \dfrac{1}{x^2} = 0.$

293. $y' = \dfrac{1}{2x - y^2}.$

294. $x^2 + xy' = 3x + y'.$

295. $4x^3y^2\,dx + (x^4 - 2x^4y - 1)\,dy = 0.$

296. $xyy' - y^2 = x^4.$

297. $\dfrac{dx}{x^2 - xy + y^2} = \dfrac{dy}{2y^2 - xy}.$

298. $(2x - 1)y' - 2y = \dfrac{1 - 4x}{x^2}.$

299. $(x - y + 3)\,dx + (3x + y + 1)\,dy = 0.$

300. $y' + \cos \dfrac{x + y}{2} = \cos \dfrac{x - y}{2}.$

301. $y'(3x^2 - 2x) - y(6x - 2) + \dfrac{2}{x}(9x - 4) = 0.$

302. $xy^2y' - y^3 = \frac{1}{3}x^4$.

303. $y' = \tan(ax + by + c)$.

304. $\left(1 + e^{\frac{x}{y}}\right) dx + e^{\frac{x}{y}}\left(1 - \frac{x}{y}\right) dy = 0, \quad y|_{x=1} = 1$.

305. $(x^2 + y^2)\,dx - xy\,dy = 0$.

306. $x[ay^3 + (ay + bx)^3]\,dx + y[bx^3 + (ay + bx)^3]\,dy = 0$.

307. $(xy^2 + y)\,dx - x\,dy = 0$.

308. $(x^2 + y^2 + 2x)\,dx + 2y\,dy = 0$.

309. $(x - 1)(y^2 - y + 1)\,dx = (y + 1)(x^2 + x + 1)\,dy$.

310. $(x - 2xy - y^2)y' + y^2 = 0$.

311. $y\cos x\,dx + (2y - \sin x)\,dy = 0$.

312. $y' - 1 = e^{x+2y}$.

313. $2(x^5 + 2x^3y - y^2x)\,dx + (y^2 + 2x^2y - x^4)\,dy = 0$.

314. $x^2y^ny' = 2xy' - y$.

315. $\{\sqrt{1 + x^2} + ny\}\,dx + \{\sqrt{1 + y^2} + nx\}\,dy = 0, \quad y|_{x=0} = n$.

316. $[3(x + y) + a^2]y' = 4(x + y) + b^2$.

317. $axyy'^2 + (x^2 - ay^2 - b)y' - xy = 0$,
(substitute $x^2 = s$, $y^2 = t$).

§11. DIFFERENTIAL EQUATIONS OF HIGHER ORDERS. REDUCTION OF THE ORDER OF AN EQUATION

A differential equation of the nth order, not solved for the highest derivative $y^{(n)}$, has the form

$$F(x, y, y', y'', \ldots, y^{(n)}) = 0 \tag{1}$$

or, if it is solved for $y^{(n)}$,

$$y^{(n)} = f(x, y, y', y'', \ldots, y^{(n-1)}). \tag{2}$$

The following theorem, on the existence and uniqueness of the solution of the differential equation (2), is valid.

Theorem. If in equation (2) the function $f(x, y, y', y'', \ldots, y^{(n-1)})$ satisfies the following conditions:

(*a*) it is continuous with respect to all its arguments

$$x, y, y', y'', \ldots, y^{(n-1)}$$

in some region D,

(*b*) it has continuous partial derivatives with respect to

$$y, y', y'', \ldots, y^{(n-1)}$$

in the region D, then there exists a unique solution $y = \phi(x)$ of equation (2), satisfying the conditions

$$y|_{x=x_0} = y_0, \; y'|_{x=x_0} = y_0', \ldots, y^{(n-1)}|_{x=x_0} = y_0^{(n-1)}, \tag{3}$$

where the values $x = x_0, y = y_0, y' = y_0' \ldots, y^{(n-1)} = y_0^{(n-1)}$ are contained in the region D.

The conditions

$$y|_{x=x_0} = y_0, \; y'|_{x=x_0} = y_0', \ldots, y^{(n-1)}|_{x=x_0} = y_0^{(n-1)}$$

are called initial conditions.

The problem of finding a solution $y = \phi(x)$ of equation (2), satisfying the initial conditions (3), is called the Cauchy problem for equation (2).

Note. For an equation of the second order $y'' = f(x, y, y')$ the initial conditions have the form

$$y|_{x=x_0} = y_0, \quad y'|_{x=x_0} = y_0',$$

where x_0, y_0, y_0' are given numbers. In this case the theorem of existence and uniqueness means geometrically that through a given point $M_0(x_0, y_0)$ of the plane XOY with given tangent of gradient y_0' there passes a unique curve.

The general solution of a differential equation of the nth order (2) means the set of all solutions determined by the formula

$$y = \phi(x, C_1, C_2, \ldots, C_n),$$

containing n arbitrary constants $C_1, C_2, \ldots, C_n$ such that if the initial conditions (3) are given, values $\tilde{C}_1, \tilde{C}_2, \ldots, \tilde{C}_n$ can be found such that $y = \phi(x, \tilde{C}_1, \tilde{C}_2, \ldots, \tilde{C}_n)$ is a solution of equation (2), satisfying these initial conditions.

Any solution obtained from the general solution for particular values of the arbitrary constants $C_1, C_2, \ldots, C_n$ is called a particular solution of the differential equation.

An equation of the form

$$\Phi(x, y, C_1, C_2, \ldots, C_n) = 0,$$

which determines implicitly the general solution of the differential equation, is called the *general integral of the equation.*

Any equation obtained from the general integral for particular values of the arbitrary constants $C_1, C_2, \ldots, C_n$ is called a *particular integral of the equation.*

The graph of a particular solution or a particular integral is called an integral curve of the given differential equation.

Example. Show that $y = C_1x + C_2$ is the general solution of the differential equation $y'' = 0$.

Solution. We show that $y = C_1x + C_2$ satisfies the given equation. In fact, we have $y' = C_1$, $y'' = 0$.

Now let arbitrary initial conditions $y|_{x=x_0} = y_0$, $y'|_{x=x_0} = y_0'$ be given. We show that the constants C_1 and C_2 can be chosen so that $y = C_1x + C_2$ satisfies these conditions. We have

$$y = C_1x + C_2, \quad y' = C_1.$$

Putting $x = x_0$, we obtain the system:

$$\left.\begin{aligned} y_0 &= C_1x_0 + C_2 \\ y_0' &= C_1 \end{aligned}\right\}.$$

This gives uniquely $C_1 = y_0'$ and $C_2 = y_0 - x_0y_0'$. Thus the solution $y = y_0'(x - x_0) + y_0$ satisfies the initial conditions. Geometrically this means that a unique line passes through any point $M_0(x_0, y_0)$ of the plane XOY with given gradient.

One initial condition, for example $y|_{x=x_0} = y_0$, obviously defines a pencil of lines with vertex at the point $M_0(x_0, y_0)$, i.e. one initial condition is insufficient to isolate a unique solution.

In the following problems, show that the given functions are solutions of the corresponding equations:

318. $y = e^{-x}(3\cos x - 2\sin x), \quad y'' + 2y' + 2y = 0.$

319. $y = e^{2x}(C_1\cos 2x + C_2\sin 2x), \quad y'' - 4y' + 8y = 0.$

320. $y = x(\sin x - \cos x), \quad y'' + y = 2(\cos x + \sin x).$

321. $y = (C_1 + C_2x)e^{-3x}, \quad y'' + 6y' + 9y = 0.$

322. $y = x^2 \ln x, \quad xy''' = 2.$

323. $x = y^2 + y, \quad y'y''' = 3y''^2.$

324. $x + C = e^{-y}, \quad y'' = y'^2.$

325. $x = y + \ln y, \quad yy'' + y'^3 - y'^2 = 0.$

326. $y = C_1 + C_2 \int_1^x \frac{e^t}{t}\, dt, \quad xy'' + (1 - x)y' = 0.$

327. $y = C_1 x + C_2 x \int_x^2 \frac{e^t}{t}\, dt \; (x > 0),$

$$x^2 y'' - (x^2 + x)y' + (x + 1)y = 0.$$

328. $y = C_1 \ln x + C_2 \ln x \int_x^e \frac{dt}{\ln t} \; (x > 1),$

$$y'' x^2 \ln^2 x - y' x \ln x + y(\ln x + 1) = 0.$$

329. $\left.\begin{array}{l} x = t(2 \ln t - 1) + C_1 \\ y = t^2 \ln t + C_2 \end{array}\right\}, \quad y''(1 + 2 \ln y') = 1.$

330. $\left.\begin{array}{l} x = (t + 1)e^t + C_1 \\ y = t^2 e^t + C_2 \end{array}\right\}, \quad y'' e^{y'}(y' + 2) = 1.$

331. $\left.\begin{array}{l} x = C_2 + C_1(t - \frac{1}{2} \sin 2t) \\ y = 1 - C_1 \sin^2 t \end{array}\right\}, \quad 2(1 - y)y'' = 1 + y'^2.$

332. $\left.\begin{array}{l} x = \dfrac{1}{2} \ln t + \dfrac{3}{4t^2} \\ y = \dfrac{1}{4} t + \dfrac{3}{4t^3} \end{array}\right\}, \quad y''^2 - 2y'y'' + 3 = 0.$

Show that the given functions are the general solutions of the corresponding equations:

333. $y = C_1 \sin x + C_2 \cos x, \quad y'' + y = 0.$

334. $y = \frac{1}{x}(C_1 e^x + C_2 e^{-x}), \quad xy'' + 2y' - xy = 0.$

335. $y = C_1 x + C_2 \ln x, \quad x^2(1 - \ln x)y'' + xy' - y = 0.$

336. $y = \sqrt{(x + C_1)^2 + C_2}, \quad yy'' + y'^2 = 1.$

337. $x + C_2 = y^3 + C_1 y, \quad y'' + 6yy'^3 = 0.$

338. $x + C_2 = \ln \sin (y + C_1), \quad y'' = y'(1 + y'^2).$

339. $y = C_1x + C_2x\int_0^x \frac{\sin t}{t^2}\,dt,\quad y''x\sin x - y'x\cos x + y\cos x = 0.$

In the following problems, show that the given relations are integrals (general or particular) of the corresponding equations:

340. $(x - C_1)^2 + (y - C_2)^2 = 1,\quad y'' = (1 + y'^2)^{3/2}.$

341. $y^2 = 1 + (1 - x)^2,\quad y^3y'' = 1.$

342. $\sin(y - C_2) = e^{x-C_1},\quad y'' = y'(1 + y'^2).$

343. $C_1x + C_2 = \ln(C_1y - 1),\quad yy'' = y'^2 + y'.$

344. $y\ln y = x + \int_0^x e^{t^2}\,dt,\quad y(1 + \ln y)y'' + y'^2 = 2xye^{x^2}.$

REDUCTION OF THE ORDER OF AN EQUATION

We shall indicate some of the most frequently occurring classes of differential equations whose order can be reduced.

I. $y^{(n)} = f(x)$.

After n integrations we obtain the general solution

$$y = \underbrace{\int\cdots\int}_{n} f(x)\,\underbrace{dx\ldots dx}_{n} + C_1\frac{x^{n-1}}{(n-1)!} + C_2\frac{x^{n-2}}{(n-2)!} + \cdots + C_{n-1}x + C_n.$$

II. The equation does not contain the unknown function and its derivatives up to order $k - 1$ inclusive:

$$F(x, y^{(k)}, \; y^{(k+1)}, \ldots, y^{(n)}) = 0.$$

The order of the equation can be reduced by the change of variables $y^{(k)}(x) = p(x)$. Then equation (1) takes the form

$$F(x, p, p', \ldots, p^{(n-k)}) = 0.$$

From this equation we determine, if possible,

$$p = f(x, C_1, C_2, \ldots, C_{n-k}),$$

and then find y from the equation $y^{(k)} = f(x, C_1, C_2, \ldots, C_{n-k})$ by integrating k times.

III. The equation does not contain the independent variable:

$$F(y, y', y'', \ldots, y^{(n)}) = 0.$$

The substitution $y' = p$ allows the order of the equation to be reduced by one. For p can be regarded as a new unknown function of y: $p = p(y)$. We express all the derivatives $y', y'', \ldots, y^{(n)}$ in terms of derivatives of the new unknown function p of y:

$$y' = \frac{dy}{dx} = p,$$

$$y'' = \frac{dp}{dx} = \frac{dp}{dy}\frac{dy}{dx} = p\frac{dp}{dy},$$

$$y''' = \frac{d}{dx}\left(p\frac{dp}{dy}\right) = \frac{d}{dy}\left(p\frac{dp}{dy}\right)\frac{dy}{dx} = p^2\frac{d^2p}{dy^2} + p\left(\frac{dp}{dy}\right)^2 \text{ etc.}$$

Substituting these expressions for $y', y'', \ldots, y^{(n)}$ into the equation, we obtain a differential equation of order $n - 1$.

IV. The equation

$$F(x, y, y', \ldots, y^{(n)}) = 0$$

is homogeneous in the arguments $y, y', y'', \ldots, y^{(n)}$, i.e.

$$F(x, ty, ty', \ldots, ty^{(n)}) = t^k F(x, y, y', \ldots, y^{(n)}).$$

The order of such an equation can be reduced by one by means of the substitution

$$y = e^{\int z\,dx},$$

where z is a new unknown function of x: $z = z(x)$.

V. An equation

$$F(x, y, dx, d^2y, \ldots, d^ny) = 0,$$

i.e. an equation written in differentials, in which the function F is homogeneous in its arguments $x, y, dx, dy, d^2y, \ldots, d^ny$, if x and dx are reckoned as being of the first degree and y, dy, d^2y etc. are reckoned as being of degree m. Then $\frac{dy}{dx}$ will have degree $m - 1$, $\frac{d^2y}{dx^2}$ will have degree $m - 2$, etc.

To reduce the order we apply the substitution

$$x = e^t, \quad y = ue^{mt}.$$

As a result we obtain a differential equation in u and t not containing t explicitly, i.e. allowing a reduction of the order by one (case III).

To solve the Cauchy problem for equations of higher order, it is advisable to determine the values of the constants C_i in the course of the solution, rather than find them from the general solution of the equation. This speeds up the solution of the problem and, in addition, it can happen that the integration is considerably simplified when the constants C_i take actual numerical values, while for arbitrary C_i the integration is difficult, and in general impossible in terms of elementary functions.

Example. Solve the equation

$$y'' = 2y^3, \quad y|_{x=0} = 1, \quad y'|_{x=0} = 1.$$

Solution. We have

$$p\frac{dp}{dy} = 2y^3, \quad \text{hence} \quad p^2 = y^4 + C_1 \quad \text{or} \quad \frac{dy}{dx} = \sqrt{y^4 + C_1}.$$

Separating the variables, we arrive at the integral $\int (y^4 + C_1)^{-1/2}\, dy$. This integral cannot be expressed as a finite combination of elementary functions. However, if we apply the initial conditions, we obtain $C_1 = 0$. So $\frac{dy}{dx} = y^2$ and hence, taking the initial conditions into account, we have finally $y = \frac{1}{1-x}$.

We shall consider examples of various cases of the reduction of the order of a differential equation.

Example. Find the general solution of the equation

$$y''' = \sin x + \cos x.$$

Solution. Integrating the given equation successively, we have

$$y'' = -\cos x + \sin x + C_1,$$

$$y' = -\sin x - \cos x + C_1 x + C_2,$$

$$y = \cos x - \sin x + C_1\frac{x^2}{2} + C_2 x + C_3.$$

Example. Find the general solution of the equation

$$y''' = \frac{\ln x}{x^2}$$

and select the solution which satisfies the initial conditions

$$y|_{x=1} = 0, \quad y'|_{x=1} = 1, \quad y''|_{x=1} = 2.$$

Solution. Integrating this equation three times, we find that

$$y'' = \int \frac{\ln x}{x^2}\,dx = -\frac{\ln x}{x} - \frac{1}{x} + C_1,$$

$$y' = -\tfrac{1}{2}\ln^2 x - \ln x + C_1 x + C_2, \qquad (1)$$

$$y = -\frac{x}{2}\ln^2 x + C_1\frac{x^2}{2} + C_2 x + C_3.$$

We find the solution which satisfies the given initial conditions. Substituting

$$y|_{x=1} = 0, \quad y'|_{x=1} = 1, \quad y''|_{x=1} = 2$$

into (1), we have

$$\left.\begin{aligned} \frac{C_1}{2} + C_2 + C_3 &= 0 \\ C_1 + C_2 &= 1 \\ -1 + C_1 &= 2 \end{aligned}\right\}.$$

Hence $C_1 = 3$, $C_2 = -2$, $C_3 = \frac{1}{2}$. The required solution is

$$y = -\frac{x}{2}\ln^2 x + \frac{3}{2}x^2 - 2x + \frac{1}{2}.$$

Example. Solve the equation

$$y''' = \sqrt{1 + (y'')^2}.$$

Solution. The given equation does not contain the function y and its first derivative, so we put $y'' = p$. Then the equation takes the form

$$\frac{dp}{dx} = \sqrt{1 + p^2}.$$

Separating the variables and integrating, we obtain

$$p = \frac{e^{x+C_1} - e^{-(x+C_1)}}{2}.$$

Replacing p by y'', we obtain

$$y'' = \frac{e^{x+C_1} - e^{-(x+C_1)}}{2}.$$

Integrating successively, we have

$$y' = \frac{e^{x+C_1} + e^{-(x+C_1)}}{2} + C_2$$

and

$$y = \frac{e^{x+C_1} - e^{-(x+C_1)}}{2} + C_2x + C_3$$

or

$$y = \operatorname{sh}(x + C_1) + C_2x + C_3.$$

Example. Solve the equation

$$xy^{\mathrm{V}} - y^{\mathrm{IV}} = 0.$$

Solution. The equation does not contain the function y and its derivatives up to the third order inclusive. So, putting $y^{\mathrm{IV}} = p$, we obtain

$$x\frac{dp}{dx} - p = 0.$$

Hence

$$p = C_1x, \quad y^{\mathrm{IV}} = C_1x.$$

Integrating successively, we find that

$$y''' = C_1\frac{x^2}{2} + C_2,$$

$$y'' = C_1\frac{x^3}{6} + C_2x + C_3,$$

$$y' = C_1\frac{x^4}{24} + C_2\frac{x^2}{2} + C_3x + C_4,$$

$$y = C_1\frac{x^5}{120} + C_2\frac{x^3}{6} + C_3\frac{x^2}{2} + C_4x + C_5$$

or

$$y = \bar{C}_1x^5 + \bar{C}_2x^3 + \bar{C}_3x^2 + C_4x + C_5,$$

where

$$\bar{C}_1 = \frac{C_1}{120}, \quad \bar{C}_2 = \frac{C_2}{6}, \quad \bar{C}_3 = \frac{C_3}{2}.$$

Example. Solve the equation

$$y'' + y'^2 = 2e^{-y}.$$

Solution. The equation does not contain the independent variable x, so putting $y' = p$, $y'' = p\,dp/dy$ into the given equation we obtain a Bernoulli equation

$$p\frac{dp}{dy} + p^2 = 2e^{-y}.$$

The substitution $p^2 = z$ reduces it to a linear equation

$$\frac{dz}{dy} + 2z = 4e^{-y}.$$

Using the method of variation of parameters, we obtain the general solution

$$z = 4e^{-y} + C_1e^{-2y}.$$

Replacing z by $p^2 = y'^2$, we obtain

$$\frac{dy}{dx} = \pm\sqrt{4e^{-y} + C_1e^{-2y}}.$$

Separating the variables and integrating, we have

$$x + C_2 = \pm\tfrac{1}{2}\sqrt{4e^{y} + C_1}.$$

Hence

$$e^y + \bar{C}_1 = (x + C_2)^2, \quad \text{where} \quad \bar{C}_1 = \frac{C_1}{4}.$$

This is the general integral of the given equation.

Example. Solve the equation

$$x^2yy'' = (y - xy')^2.$$

Solution. The given equation is homogeneous in y, y', y''. The order of this equation is reduced by one by means of the substitution $y = e^{\int z\,dx}$, where z is a new unknown function of x. We have

$$y' = ze^{\int z\,dx}, \quad y'' = (z' + z^2)e^{\int z\,dx}.$$

Substituting the expressions for y, y', y'' into the equation, we obtain

$$x^2(z' + z^2)e^{2\int z\,dz} = (e^{\int z\,dx} - xze^{\int z\,dx})^2.$$

We cancel $e^{2\int z\,dx}$:

$$x^2(z' + z^2) = (1 - xz)^2 \quad \text{or} \quad x^2z' + 2xz = 1.$$

This equation is linear. Its left hand side can be written in the form

$$(x^2z)' = 1.$$

Hence

$$x^2z = x + C_1$$

or

$$z = \frac{1}{x} + \frac{C_1}{x^2}.$$

We find the integral

$$\int z\,dx = \int\left(\frac{1}{x} + \frac{C_1}{x^2}\right)dx = \ln|x| - \frac{C_1}{x} + \ln C_2.$$

The general solution of the given equation is

$$y = e^{\ln|x| - \frac{C_1}{x} + \ln C_2} \quad \text{or} \quad y = C_2xe^{-\frac{C_1}{x}}.$$

In addition, the equation has the obvious solution $y \equiv 0$.

Example. Solve the equation

$$x^3y'' = (y - xy')^2.$$

Solution. We show that this equation is generalized homogeneous, i.e. we find the number m. Equating the degrees of all the terms, counting x, y, y', y'' as quantities of degrees $1, m, m - 1$ and $m - 2$ respectively, we have

$$3 + m - 2 = 2m,$$

hence $m = 1$. We now make the substitution

$$x = e^t, \quad y = ue^t.$$

Since

$$\frac{dy}{dx} = \frac{\dfrac{dy}{dt}}{\dfrac{dx}{dt}} = \frac{\left(\dfrac{du}{dt} + u\right)e^t}{e^t} = \frac{du}{dt} + u,$$

$$\frac{d^2y}{dx^2} = \frac{\dfrac{d}{dt}\left(\dfrac{dy}{dx}\right)}{\dfrac{dx}{dt}} = \frac{\dfrac{d^2u}{dt^2} + \dfrac{du}{dt}}{e^t} = e^{-t}\left(\frac{d^2u}{dt^2} + \frac{du}{dt}\right),$$

the given equation, after cancellation of e^{2t}, takes the form

$$\frac{d^2u}{dt^2} + \frac{du}{dt} = \left(\frac{du}{dt}\right)^2.$$

Putting $\frac{du}{dt} = p$, $\frac{d^2u}{dt^2} = p\frac{dp}{du}$, we obtain

$$p\frac{dp}{du} + p = p^2.$$

Hence

$$p = 0 \quad \text{or} \quad \frac{dp}{du} + 1 = p.$$

Integrating the second equation, we find that

$$p = 1 + C_1e^u \quad \text{or} \quad \frac{du}{dt} = 1 + C_1e^u.$$

The general solution of this equation is

$$u = \ln\frac{e^t}{C_1e^t + C_2}.$$

Returning to the variables x and y, we obtain the general solution of the given equation

$$y = x\ln\frac{x}{C_1x + C_2}.$$

Note. The case $p = 0$ gives $u = C$ or $y = Cx$, a particular solution obtained from the general solution when

$$C_1 = e^{-C}, \quad C_2 = 0.$$

Integrate the equations:

345. $y''' = xe^x$, $y(0) = y'(0) = y''(0) = 0.$

346. $y^{\text{IV}} = x.$

347. $y''' = x\ln x$, $y(1) = y'(1) = y''(1) = 0.$

348. $y''' = x + \cos x.$

349. $y''' = \frac{x}{(x+2)^5}$, $y(1) = y'(1) = y''(1) = 0.$

350. $y''^2 - 5y' + 6 = 0.$

351. $(1 + x^2)y'' + y'^2 + 1 = 0.$

352. $y''^2 - 2y''y' + 3 = 0.$

353. $xy'' = y' \ln \frac{y'}{x}.$

354. $y''^2 + y'^2 = y'^4.$

355. $y''^2 + y'''^2 = 1.$

356. $y''(1 + 2 \ln y') = 1.$

357. $x = y''^2 + 1.$

358. $4y' + y''^2 = 4xy''.$

359. $y''^2 - y'y''' = \left(\frac{y'}{x}\right)^2.$

360. $y''(y' + 2)e^{y'} = 1.$

361. $y'' = \frac{y'}{x} + \frac{x^2}{y'}, \quad y(2) = 0, \quad y'(2) = 4.$

362. $xy''' + y'' - x - 1 = 0.$

363. $y'y''' - 3y''^2 = 0.$

364. $xy'^2y'' - y'^3 = \frac{1}{3}x^4.$

365. $x^4y''' + 2x^3y'' = 1.$

366. $\sqrt{1 - x^2}\, y'' + \sqrt{1 - y'^2} = 0.$

367. $(x - 1)y''' + 2y'' = \frac{x + 1}{2x^2}.$

368. $y''y^3 = 1.$

369. $yy'' - y'^2 - 1 = 0.$

370. $2y''y^2 = 1.$

371. $2y'' = ae^y.$

372. $y'' = \frac{1}{4\sqrt{y}}.$

373. $3y'' = y^{-5/3}.$

374. $1 + y'^2 = 2yy''.$

375. $y^3y'' = -1, \quad y(1) = 1, \quad y'(1) = 0.$

376. $y^4 - y^3y'' = 1.$

377. $yy'' - y'^2 = y^2y'.$

378. $yy'' = y'^2.$

379. $y'' = e^{2y}, \quad y(0) = 0, \quad y'(0) = 1.$

380. $2yy'' - 3y'^2 = 4y^2.$

381. $y'' = (1 + y'^2)^{3/2}.$

382. $xy'(yy'' - y'^2) - yy'^2 = x^4y^3.$

383. $x^4y'' = (y - xy')^3, \quad y(1) = y'(1) = 1.$

384. $y^2y''' - 3yy'y'' + 2y'^3 + \frac{y}{x}(yy'' - y'^2) = \frac{y^3}{x^2}.$

385. Find the time taken for a body to fall to the earth from a height of 400,000 km (the approximate distance of the moon from the centre of the earth), if its height is measured from the centre of the earth and if the radius of the earth is approximately 6,400 km.
386. Find the law of motion of a particle of mass m on a straight line OA under the influence of a repulsive force inversely proportional to the cube of the distance $x = OM$ of the particle from the fixed centre O.
387. A body of mass m falls from a certain height. During its fall, the body undergoes a resistance proportional to the square of the speed. Find the law of motion of the falling body.
388. Find a curve passing through the origin and such that the area of the triangle formed by the tangent to the curve at any point, the ordinate of this point and the axis OX is proportional to the area of the curvilinear triangle formed by the curve, the axis OX, and the ordinate of this point.
389. Find a curve whose radius of curvature is constant.

§12. LINEAR DIFFERENTIAL EQUATIONS OF THE n-th ORDER

1. Linear independence of functions. The Wronskian.

Suppose we have a finite system of n functions

$$y_1(x), y_2(x), \ldots \ y_n(x),$$

defined on the interval (a, b). The functions $y_1(x), y_2(x), \ldots, y_n(x)$ are called linearly dependent on the interval (a, b) if there exist constants $\alpha_1, \alpha_2, \ldots, \alpha_n$, not all zero, such that for all values of x in this interval the identity

$$\alpha_1 y_1(x) + \alpha_2 y_2(x) + \ldots + \alpha_n y_n(x) \equiv 0$$

holds. If this identity is satisfied only if $\alpha_1 = \alpha_2 = \ldots = \alpha_n = 0$, the functions $y_1(x), y_2(x), \ldots, y_n(x)$ are called linearly independent on the interval (a, b).

Example 1. The system of functions $1, x, x^2, x^3$ is linearly independent on the interval $(-\infty, +\infty)$. In fact, the equation

$$\alpha_1 \cdot 1 + \alpha_2 x + \alpha_3 x^2 + \alpha_4 x^3 = 0$$

can be satisfied for all x in $(-\infty, +\infty)$ only on condition that $\alpha_1 = \alpha_2 = \alpha_3 = \alpha_4 = 0$. If at least one of these numbers is not zero, then the left hand side of the equation will be a polynomial of degree not greater than three, and it can be equal to zero for no more than three values of x in the given interval.

Example 2. The system of functions $e^{k_1 x}, e^{k_2 x}, e^{k_3 x}$, where k_1, k_2, k_3 are all different, is linearly independent on the interval $-\infty < x < +\infty$.

Suppose, conversely, that the given system of functions is linearly dependent on this interval. Then

$$\alpha_1 e^{k_1 x} + \alpha_2 e^{k_2 x} + \alpha_3 e^{k_3 x} \equiv 0 \tag{A}$$

on the interval $(-\infty, +\infty)$, where at least one of the numbers $\alpha_1, \alpha_2, \alpha_3$ is non-zero, say $\alpha_3 \neq 0$. Dividing both sides of the identity (A) by $e^{k_1 x}$, we have

$$\alpha_1 + \alpha_2 e^{(k_2 - k_1)x} + \alpha_3 e^{(k_3 - k_1)x} \equiv 0. \tag{1}$$

Differentiating (1), we obtain

$$\alpha_2(k_2 - k_1)e^{(k_2 - k_1)x} + \alpha_3(k_3 - k_1)e^{(k_3 - k_1)x} \equiv 0. \tag{2}$$

We divide both sides of the identity (2) by $e^{(k_2 - k_1)x}$:

$$\alpha_2(k_2 - k_1) + \alpha_3(k_3 - k_1)e^{(k_3 - k_2)x} \equiv 0. \tag{3}$$

Differentiating (3), we obtain

$$\alpha_3(k_3 - k_1)(k_3 - k_2)e^{(k_3 - k_2)x} \equiv 0,$$

which is impossible, since $\alpha_3 \neq 0$ by hypothesis, $k_3 \neq k_1$ and $k_3 \neq k_2$ (given), and $e^{(k_3-k_2)x} \neq 0$.

Our assumption about the linear dependence of the given system of functions has led to a contradiction, and so this system of functions is linearly independent on the interval $(-\infty, +\infty)$, i.e. the identity (A) is satisfied only if $\alpha_1 = \alpha_2 = \alpha_3 = 0$.

Example 3. The system of functions $e^{\alpha x} \sin \beta x$, $e^{\alpha x} \cos \beta x$, where $\beta \neq 0$, is linearly independent on the interval $-\infty < x < +\infty$.

We determine the values of α_1 and α_2 for which the identity

$$\alpha_1 e^{\alpha x} \sin \beta x + \alpha_2 e^{\alpha x} \cos \beta x \equiv 0. \tag{1}$$

is satisfied. Dividing both sides by $e^{\alpha x} \neq 0$, we have

$$\alpha_1 \sin \beta x + \alpha_2 \cos \beta x \equiv 0. \tag{2}$$

Substituting into (2) the value $x = 0$, we obtain $\alpha_2 = 0$, which means that $\alpha_1 \sin \beta x \equiv 0$; but the function $\sin \beta x$ is not identically zero, so $\alpha_1 = 0$. The identity (2), and consequently the identity (1), holds only if $\alpha_1 = \alpha_2 = 0$, i.e. the given functions are linearly independent on the interval $-\infty < x < +\infty$.

Note. We have incidentally proved the linear independence of the trigonometric functions $\sin \beta x$, $\cos \beta x$.

Example 4. The functions $\sin x$, $\sin (x + \frac{1}{8}\pi)$, $\sin (x - \frac{1}{8}\pi)$ are linearly dependent on the interval $(-\infty, +\infty)$.

We shall show that there exist numbers α_1, α_2, α_3, not all zero, such that on the interval the identity

$$\alpha_1 \sin x + \alpha_2 \sin \left(x + \frac{\pi}{8}\right) + \alpha_3 \sin \left(x - \frac{\pi}{8}\right) \equiv 0 \tag{A}$$

holds. Assuming that identity (A) is satisfied, we put, for example $x = 0$, $x = \frac{1}{4}\pi$, $x = \frac{1}{2}\pi$. Then we obtain a homogeneous system of three equations in three unknowns α_1, α_2, α_3:

$$\left.\begin{aligned} \alpha_2 \sin \frac{\pi}{8} - \alpha_3 \sin \frac{\pi}{8} &= 0 \\ \alpha_1 \frac{1}{\sqrt{2}} + \alpha_2 \sin \frac{3\pi}{8} + \alpha_3 \sin \frac{\pi}{8} &= 0 \\ \alpha_1 + \alpha_2 \sin \frac{5\pi}{8} + \alpha_3 \sin \frac{3\pi}{8} &= 0 \end{aligned}\right\}. \tag{1}$$

The determinant of this system is

$$\Delta = \begin{vmatrix} 0 & \sin\frac{\pi}{8} & -\sin\frac{\pi}{8} \\ \frac{1}{\sqrt{2}} & \sin\frac{3\pi}{8} & \sin\frac{\pi}{8} \\ 1 & \sin\frac{5\pi}{8} & \sin\frac{3\pi}{8} \end{vmatrix} = 0.$$

Consequently the homogeneous system (1) has non-zero solutions, i.e. there exist numbers α_1, α_2, α_3, of which at least one is non-zero. To find such a set of numbers α_1, α_2, α_3, we choose, for example, the first two equations of system (1):

$$\left.\begin{aligned} \alpha_2 \sin\frac{\pi}{8} - \alpha_3 \sin\frac{\pi}{8} &= 0 \\ \frac{\alpha_1}{\sqrt{2}} + \alpha_2 \sin\frac{3}{8}\pi + \alpha_3 \sin\frac{\pi}{8} &= 0 \end{aligned}\right\}$$

From the first equation we have $\alpha_2 = \alpha_3$. From the second, we have

$$\alpha_1 = -\sqrt{2}\left(\sin\frac{3}{8}\pi + \sin\frac{\pi}{8}\right)\alpha_3$$

or

$$\alpha_1 = -2\cos\frac{\pi}{8}\,.\,\alpha_3.$$

Putting $\alpha_3 = 1$, we obtain a non-zero solution of the system (1):

$$\alpha_1 = -2\cos\frac{\pi}{8}, \quad \alpha_2 = 1, \quad \alpha_3 = 1.$$

We show now that for these values of α_1, α_2, α_3 the identity (A) is satisfied for all x in $(-\infty, +\infty)$. In fact, we have

$$\begin{aligned} &\alpha_1 \sin x + \alpha_2 \sin\left(x + \frac{\pi}{8}\right) + \alpha_3 \sin\left(x - \frac{\pi}{8}\right) \\ &\equiv -2\cos\frac{\pi}{8}\sin x + \sin\left(x + \frac{\pi}{8}\right) + \sin\left(x - \frac{\pi}{8}\right) \\ &\equiv -2\cos\frac{\pi}{8}\sin x + 2\sin x\cos\frac{\pi}{8} \equiv 0, \end{aligned}$$

whatever the value of x. Consequently the given system of functions is linearly dependent on the interval $-\infty < x < +\infty$.

Note. In the case of two functions it is possible to give a simpler criterion for linear independence. Namely, the functions $\phi_1(x)$ and $\phi_2(x)$ are linearly independent on the interval (a, b) if their ratio is not identically equal to a constant: $\dfrac{\phi_1(x)}{\phi_2(x)} \not\equiv$ const. on this interval; if $\dfrac{\phi_1(x)}{\phi_2(x)} \equiv$ const., then the functions are linearly dependent.

Example 1. The functions $\tan x$ and $\cot x$ are linearly independent on the interval $0 < x < \frac{1}{2}\pi$, since their ratio $\dfrac{\tan x}{\cot x} = \tan^2 x \not\equiv$ const. on this interval.

Example 2. The functions $\sin 2x$ and $\sin x \cos x$ are linearly dependent on the interval $-\infty < x < +\infty$, since their ratio $\dfrac{\sin 2x}{\sin x \cos x} \equiv \dfrac{2 \sin x \cos x}{\sin x \cos x} \equiv 2$ (at points where the function $\dfrac{\sin 2x}{\sin x \cos x}$ is undefined, we define this ratio by continuity).

Investigate whether the given functions are linearly independent.

390. $4, x$.

391. $1, 2, x, x^2$.

392. $x, 2x, x^2$.

393. e^x, xe^x, x^2e^x.

394. $\sin x, \cos x, \cos 2x$.

395. $1, \sin x, \cos 2x$.

396. $5, \cos^2 x, \sin^2 x$.

397. $\cos x, \cos (x + 1), \cos (x - 2)$.

398. $1, \sin 2x, (\sin x - \cos x)^2$.

399. $x, a^{\log_a x}$ $(x > 0)$.

400. $\log_a x, \log_a x^2$ $(x > 0)$.

401. $1, \text{arc} \sin x, \text{arc} \cos x$.

402. $5, \text{arc} \tan x, \text{arc} \cot x$.

403. 2π, $\arctan\frac{x}{2\pi}$, $\operatorname{arc\,cot}\frac{x}{2\pi}$.

404. $e^{-\frac{ax^2}{2}}$, $e^{-\frac{ax^2}{2}}\int_0^x e^{\frac{at^2}{2}}\,dt$.

405. x, $x\int_{x_0}^{1}\frac{e^t}{t^2}\,dt \qquad (x_0 > 0)$.

Suppose that the n functions $y_1(x), y_2(x), \ldots, y_n(x)$ have derivatives of the $(n-1)$-th order. The determinant

$$W[y_1, y_2, \ldots, y_n] = \begin{vmatrix} y_1(x) & y_2(x) \ldots & y_n(x) \\ y_1'(x) & y_2'(x) \ldots & y_n'(x) \\ \ldots & \ldots & \ldots \\ y_1^{(n-1)}(x) & y_2^{(n-1)}(x) \ldots & y_n^{(n-1)}(x) \end{vmatrix}$$

is called the Wronskian for these functions. We note that the Wronskian is, in general, a function of x, defined on a certain interval.

The Wronskian for three functions $y_1(x), y_2(x), y_3(x)$ has the form

$$W[y_1, y_2, y_3] = \begin{vmatrix} y_1(x) & y_2(x) & y_3(x) \\ y_1'(x) & y_2'(x) & y_3'(x) \\ y_1''(x) & y_2''(x) & y_3''(x) \end{vmatrix}.$$

Example. Find the Wronskian for the functions

$$y_1(x) = e^{k_1x}, \quad y_2(x) = e^{k_2x}, \quad y_3(x) = e^{k_3x}.$$

Solution. We have

$$W[y_1, y_2, y_3] = \begin{vmatrix} e^{k_1x} & e^{k_2x} & e^{k_3x} \\ k_1e^{k_1x} & k_2e^{k_2x} & k_3e^{k_3x} \\ k_1^2e^{k_1x} & k_2^2e^{k_2x} & k_3^2e^{k_3x} \end{vmatrix}$$

$$= e^{(k_1+k_2+k_3)x}(k_2 - k_1)(k_3 - k_1)(k_3 - k_2).$$

Example. Find the Wronskian for the functions

$$y_1(x) = \sin x, \quad y_2(x) = \sin\left(x + \frac{\pi}{8}\right), \quad y_3(x) = \sin\left(x - \frac{\pi}{8}\right).$$

Solution. We have

$$W[y_1, y_2, y_3] = \begin{vmatrix} \sin x & \sin\left(x+\frac{\pi}{8}\right) & \sin\left(x-\frac{\pi}{8}\right) \\ \cos x & \cos\left(x+\frac{\pi}{8}\right) & \cos\left(x-\frac{\pi}{8}\right) \\ -\sin x & -\sin\left(x+\frac{\pi}{8}\right) & -\sin\left(x-\frac{\pi}{8}\right) \end{vmatrix} = 0,$$

since the first and last rows of the determinant are proportional.

In the following problems, find the Wronskian for the given system of functions:

406. $1, x$.

407. $x, \dfrac{1}{x}\cdot$

408. $1, 2, x^2$.

409. e^{-x}, xe^{-x}.

410. $e^x, 2e^x, e^{-x}$.

411. $2, \cos x, \cos 2x$.

412. $\sin x, \sin\left(x+\dfrac{\pi}{4}\right)\cdot$

413. $\operatorname{arc\,cos} \dfrac{x}{\pi}, \operatorname{arc\,sin} \dfrac{x}{\pi}\cdot$

414. $\pi, \operatorname{arc\,sin} x, \operatorname{arc\,cos} x$.

415. $4, \sin^2 x, \cos 2x$.

416. $x, \ln x$.

417. $\dfrac{1}{x}, e^{\frac{1}{x}}$.

418. $e^x \sin x, e^x \cos x$.

419. $e^{-3x} \sin 2x, e^{-3x} \cos 2x$.

420. $\cos x, \sin x$.

421. $\sin\left(\dfrac{\pi}{4}-x\right), \cos\left(\dfrac{\pi}{4}-x\right)\cdot$

There is a close connection between the linear dependence of a system of functions and the Wronskian of these functions, as the following theorem shows.

Theorem. If a system of functions $y_1(x)$, $y_2(x), \ldots, y_n(x)$ is linearly dependent on the interval (a, b), then the Wronskian of these functions is identically zero on this interval.

For example, we showed that the system of functions $\sin x$, $\sin(x + \frac{1}{8}\pi)$, $\sin(x - \frac{1}{8}\pi)$ is linearly dependent on the interval $(-\infty, +\infty)$, and then we verified that their Wronskian is equal to zero.

This theorem gives a necessary condition for the linear dependence of a system of functions. The converse of the theorem is untrue, i.e. the Wronskian can be equal to zero even when the given functions form a linearly independent system on some interval.

Example. Consider the two functions

$$y_1(x) = \begin{cases} 0, & \text{if } 0 \leqslant x \leqslant \frac{1}{2} \\ (x - \frac{1}{2})^2, & \text{if } \frac{1}{2} < x \leqslant 1, \end{cases}$$

$$y_2(x) = \begin{cases} (x - \frac{1}{2})^2, & \text{if } 0 \leqslant x \leqslant \frac{1}{2} \\ 0, & \text{if } \frac{1}{2} < x \leqslant 1. \end{cases}$$

Their graph has the form indicated in Fig. 14.

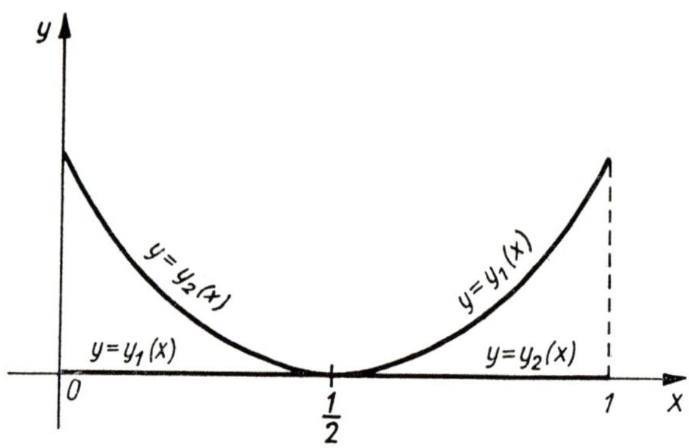

FIGURE 14

This system of functions is linearly independent, since the identity $\alpha_1 y_1(x) + \alpha_2 y_2(x) \equiv 0$ is satisfied only if $\alpha_1 = \alpha_2 = 0$. For, if we consider the interval $[0, \frac{1}{2}]$, we obtain $\alpha_2 y_2(x) \equiv 0$, so $\alpha_2 = 0$, since $y_2(x) \not\equiv 0$; on the interval $[\frac{1}{2}, 1]$ we have $\alpha_1 y_1(x) \equiv 0$, so $\alpha_1 = 0$, since $y_1(x) \not\equiv 0$ on this interval.

We find the Wronskian

$$W[y_1, y_2] = \begin{vmatrix} y_1(x) & y_2(x) \\ y_1'(x) & y_2'(x) \end{vmatrix}.$$

We have, on the interval $[0, \frac{1}{2}]$

$$W[y_1, y_2] = \begin{vmatrix} 0 & (x - \frac{1}{2})^2 \\ 0 & 2(x - \frac{1}{2}) \end{vmatrix} = 0,$$

and on the interval $[\frac{1}{2}, 1]$

$$W[y_1, y_2] = \begin{vmatrix} (x - \frac{1}{2})^2 & 0 \\ 2(x - \frac{1}{2}) & 0 \end{vmatrix} = 0.$$

Thus the Wronskian $W[y_1, y_2] \equiv 0$ on the interval [0, 1].

In the following problems, show that the given functions are linearly independent, but that their Wronskian is identically zero, and draw the graphs of these functions:

422.
$$y_1(x) = \begin{cases} x^2, & \text{if } -1 \leqslant x \leqslant 0 \\ 0, & \text{if } \quad 0 < x \leqslant 1, \end{cases}$$
$$y_2(x) = \begin{cases} 0, & \text{if } -1 \leqslant x \leqslant 0 \\ x^2, & \text{if } \quad 0 < x \leqslant 1. \end{cases}$$

423.
$$y_1(x) = \begin{cases} 0, & \text{if } 0 \leqslant x \leqslant 2 \\ (x-2)^2, & \text{if } 2 < x \leqslant 4, \end{cases}$$
$$y_2(x) = \begin{cases} (x-2)^2, & \text{if } 0 \leqslant x \leqslant 2 \\ 0, & \text{if } 2 < x \leqslant 4. \end{cases}$$

424.
$$y_1(x) = \begin{cases} x^3, & \text{if } -2 \leqslant x \leqslant 0 \\ 0, & \text{if } \quad 0 < x \leqslant 1, \end{cases}$$
$$y_2(x) = \begin{cases} 0, & \text{if } -2 \leqslant x \leqslant 0 \\ x^2, & \text{if } \quad 0 < x \leqslant 1. \end{cases}$$

425. $y_1(x) = x^2$, $y_2(x) = x|x|$; $\quad -1 \leqslant x \leqslant 1$.

We shall give yet another criterion for the linear dependence of a system of functions.

Suppose that we have a system of functions $y_1(x), y_2(x), \ldots, y_n(x)$, given on an interval $[a, b]$. We put

$$(y_i, y_j) = \int_a^b y_i(x) y_j(x)\, dx \qquad (i, j = 1, 2, \ldots, n).$$

The determinant

$$\Gamma(y_1, y_2 \ldots y_n) = \begin{vmatrix} (y_1, y_1) & (y_1, y_2) \ldots (y_1, y_n) \\ (y_2, y_1) & (y_2, y_2) \ldots (y_2, y_n) \\ (y_n, y_1) & (y_n, y_2) \ldots (y_n, y_n) \end{vmatrix}$$

is called the Gramian of the system of functions $\{y_k(x)\}$.

Theorem. In order that the system of functions $y_1(x), y_2(x), \ldots, y_n(x)$ should be linearly dependent, it is necessary and sufficient that its Gramian should be equal to zero.

Example. Show that the functions $y_1(x) = x$, $y_2(x) = 2x$ are linearly dependent on the interval [0, 1].

Solution. We have

$$(y_1, y_1) = \int_0^1 x^2\, dx = \tfrac{1}{3}; \quad (y_1, y_2) = (y_2, y_1) = \int_0^1 2x^2\, dx = \tfrac{2}{3};$$

$$(y_2, y_2) = \int_0^1 4x^2\, dx = \tfrac{4}{3}.$$

$$\Gamma(y_1, y_2) = \begin{vmatrix} \frac{1}{3} & \frac{2}{3} \\ \frac{2}{3} & \frac{4}{3} \end{vmatrix} = 0,$$

and so the functions $y_1(x)$ and $y_2(x)$ are linearly dependent. Use this criterion to show that the systems of functions in problems 392, 396, 398 are linearly dependent on $[-\pi, \pi]$.

2. Homogeneous linear equations with constant coefficients

Suppose we have a differential equation

$$a_0 y^{(n)} + a_1 y^{(n-1)} + \ldots + a_n y = 0, \tag{1}$$

where $a_0, a_1, \ldots, a_n$ are real constants.

We form the characteristic equation of equation (1):

$$a_0 \lambda^n + a_1 \lambda^{n-1} + \ldots + a_n = 0. \tag{2}$$

Let $\lambda_1, \lambda_2 \ldots, \lambda_n$ be the roots of equation (2), some of which may be repeated.

The following cases can occur:

(*a*) $\lambda_1, \lambda_2, \ldots, \lambda_n$ are real and distinct.

Then the fundamental system of solutions of equation (1) has the form:

$$e^{\lambda_1 x}, e^{\lambda_2 x}, \ldots, e^{\lambda_n x}$$

and the general solution* is

$$y_{g.h.} = C_1 e^{\lambda_1 x} + C_2 e^{\lambda_2 x} + \ldots + C_n e^{\lambda_n x}.$$

(*b*) The roots of the characteristic equation are real, but one of them is repeated. Suppose, for example, that

$$\lambda_1 = \lambda_2 = \ldots = \lambda_k = \bar{\lambda},$$

i.e. $\bar{\lambda}$ is a k-fold root of equation (2), and all the other $n - k$ roots are distinct.

The fundamental system of solutions in this case has the form

$$e^{\bar{\lambda} x}, xe^{\bar{\lambda} x}, x^2 e^{\bar{\lambda} x}, \ldots, x^{k-1} e^{\bar{\lambda} x}, e^{\lambda_{k+1} x}, \ldots, e^{\lambda_n x}$$

and the general solution is

$$y = C_1 e^{\bar{\lambda} x} + C_2 x e^{\bar{\lambda} x} + C_3 x^2 e^{\bar{\lambda} x} + \ldots + C_k x^{k-1} e^{\bar{\lambda} x} + C_{k+1} e^{\lambda_{k+1} x} + \ldots + C_n e^{\lambda_n x}.$$

(*c*) Some of the roots of the characteristic equation are complex. Suppose, for example that $\lambda_1 = \alpha + i\beta$, $\lambda_2 = \alpha - i\beta$, $\lambda_3 = \gamma + i\delta$, $\lambda_4 = \gamma - i\delta$, and the other roots are real (since by hypothesis the coefficients $a_i (i = 0, 1, 2, \ldots, n)$ of equation (1) are real, the complex roots of equation (2) are conjugate in pairs).

The fundamental system of solutions in this case has the form

$$e^{\alpha x} \cos \beta x, e^{\alpha x} \sin \beta x, e^{\gamma x} \cos \delta x, e^{\gamma x} \sin \delta x, e^{\lambda_5 x}, e^{\lambda_6 x}, \ldots, e^{\lambda_n x},$$

and the general solution appears as

$$y_{g.h.} = C_1 e^{\alpha x} \cos \beta x + C_2 e^{\alpha x} \sin \beta x + C_3 e^{\gamma x} \cos \delta x + C_4 e^{\gamma x} \sin \delta x + C_5 e^{\lambda_5 x} + \ldots + C_n e^{\lambda_n x}.$$

(*d*) in the case where $\lambda_1 = \alpha + i\beta$ is a k-fold root of equation (2) $(k \leqslant \frac{1}{2}n)$, $\lambda_2 = \alpha - i\beta$ is also a k-fold root, and the fundamental system has the form

$$e^{\alpha x} \cos \beta x, e^{\alpha x} \sin \beta x, xe^{\alpha x} \cos \beta x, xe^{\alpha x} \sin \beta x, \ldots, x^{k-1} e^{\alpha x} \cos \beta x, x^{k-1} e^{\alpha x} \sin \beta x, e^{\lambda_{2k+1} x}, \ldots, e^{\lambda_n x},$$

* The suffixes in $y_{g.h.}$ stand for general, homogeneous.

and so the general solution is

$$y_{g.h.} = C_1e^{\alpha x}\cos\beta x + C_2e^{\alpha x}\sin\beta x + C_3xe^{\alpha x}\cos\beta x$$
$$+ C_4xe^{\alpha x}\sin\beta x + \ldots + C_{2k-1}x^{k-1}e^{\alpha x}\cos\beta x$$
$$+ C_{2k}x^{k-1}e^{\alpha x}\sin\beta x + C_{2k+1}e^{\lambda_{2k+1}x} + \ldots + C_ne^{\lambda_n x}.$$

Example. Find the general solution of the equation

$$y''' - 2y'' - 3y' = 0.$$

Solution. We form the characteristic equation

$$\lambda^3 - 2\lambda^2 - 3\lambda = 0.$$

We find its roots: $\lambda_1 = 0$, $\lambda_2 = -1$, $\lambda_3 = 3$. Since they are real and distinct, the general solution has the form

$$y_{g.h.} = C_1 + C_2e^{-x} + C_3e^{3x}.$$

Example. Find the general solution of the equation

$$y''' + 2y'' + y' = 0.$$

Solution. The characteristic equation has the form

$$\lambda^3 + 2\lambda^2 + \lambda = 0.$$

Hence $\lambda_1 = \lambda_2 = -1$, $\lambda_3 = 0$. Its roots are real, but one of them, namely -1, is a double root, so the general solution has the form

$$y_{g.h.} = C_1e^{-x} + C_2xe^{-x} + C_3.$$

Example. Find the general solution of the equation

$$y''' + 4y'' + 13y' = 0.$$

Solution. The characteristic equation

$$\lambda^3 + 4\lambda^2 + 13\lambda = 0$$

has the following roots:

$$\lambda_1 = 0, \quad \lambda_2 = -2 - 3i, \quad \lambda_3 = -2 + 3i.$$

In this case the general solution is

$$y_{g.h.} = C_1 + C_2e^{-2x}\cos 3x + C_3e^{-2x}\sin 3x.$$

Example. Find the general solution of the equation

$$y^{\mathrm{V}} - 2y^{\mathrm{IV}} + 2y''' - 4y'' + y' - 2y = 0.$$

Solution. The characteristic equation

$$\lambda^5 - 2\lambda^4 + 2\lambda^3 - 4\lambda^2 + \lambda - 2 = 0$$

or

$$(\lambda - 2)(\lambda^2 + 1)^2 = 0$$

has the following roots: $\lambda = 2$ is a simple root and $\lambda = \pm i$ are double imaginary roots. The general solution is

$$y_{g.h.} = C_1 e^{2x} + (C_2 + C_3 x)\cos x + (C_4 + C_5 x)\sin x.$$

Example. Solve the equation

$$y^{\text{IV}} + 4y''' + 8y'' + 8y' + 4y = 0.$$

Solution. We form the characteristic equation

$$\lambda^4 + 4\lambda^3 + 8\lambda^2 + 8\lambda + 4 = 0 \quad \text{or} \quad (\lambda^2 + 2\lambda + 2)^2 = 0.$$

It has double complex roots:

$$\lambda_1 = \lambda_2 = -1 - i, \quad \lambda_3 = \lambda_4 = -1 + i,$$

and so the general solution has the form

$$y_{g.h.} = C_1 e^{-x}\cos x + C_2 e^{-x}\sin x + C_3 x e^{-x}\cos x + C_4 x e^{-x}\sin x$$

or

$$y_{g.h.} = e^{-x}(C_1 + C_3 x)\cos x + e^{-x}(C_2 + C_4 x)\sin x.$$

Form the homogeneous linear equations, knowing their characteristic equations:

426. $\lambda^2 + 3\lambda + 2 = 0.$

427. $2\lambda^2 - 3\lambda - 5 = 0.$

428. $\lambda(\lambda + 1)(\lambda + 2) = 0.$

429. $(\lambda^2 + 1)^2 = 0.$

430. $\lambda^3 = 0.$

Form the homogeneous linear equations if the roots of their characteristic equations are known, and write down their general solutions:

431. $\lambda_1 = 1, \quad \lambda_2 = 2.$

432. $\lambda_1 = 1, \quad \lambda_2 = 1.$

433. $\lambda_1 = 3 - 2i, \quad \lambda_2 = 3 + 2i.$

434. $\lambda_1 = 1, \quad \lambda_2 = 1, \quad \lambda_3 = 1.$

Form the homogeneous linear equations if their fundamental systems of solutions are given:

435. e^{-x}, e^{x}.

436. 1, e^{x}.

437. e^{-2x}, xe^{-2x}.

438. $\sin 3x$, $\cos 3x$.

439. 1, x.

440. e^{x}, e^{2x}, e^{3x}.

441. e^{x}, xe^{x}, x^2e^{x}.

442. e^{x}, xe^{x}, e^{2x}.

443. 1, x, e^{x}.

444. 1, $\sin x$, $\cos x$.

445. e^{2x}, $\sin x$, $\cos x$.

446. 1, $e^{-x}\sin x$, $e^{-x}\cos x$.

Integrate the following equations:

447. $y'' - y = 0$.

448. $3y'' - 2y' - 8y = 0$.

449. $y''' - 3y'' + 3y' - y = 0$, $y(0) = 1$, $y'(0) = 2$, $y''(0) = 3$.

450. $y'' + 2y' + y = 0$.

451. $y'' - 4y' + 3y = 0$. $y(0) = 6$, $y'(0) = 10$.

452. $y''' + 6y'' + 11y' + 6y = 0$.

453. $y'' - 2y' - 2y = 0$.

454. $y^{\mathrm{VI}} + 2y^{\mathrm{V}} + y^{\mathrm{IV}} = 0$.

455. $4y'' - 8y' + 5y = 0$.

456. $y''' - 8y = 0$.

457. $y^{\mathrm{IV}} + 4y''' + 10y'' + 12y' + 5y = 0$.

458. $y'' - 2y' + 2y = 0$, $y(0) = 0$, $y'(0) = 1$.

459. $y'' - 2y' + 3y = 0$, $y(0) = 1$, $y'(0) = 3$.

460. $y^{IV} + 2y''' + 4y'' - 2y' - 5y = 0.$

461. $y^{V} + 4y^{IV} + 5y''' - 6y' - 4y = 0.$

462. $y''' + 2y'' - y' - 2y = 0.$

463. $y''' - 2y'' + 2y' = 0.$

464. $y^{IV} - y = 0.$

465. $y^{X} = 0.$

466. $y''' - 3y' - 2y = 0.$

467. $2y''' - 3y'' + y' = 0.$

3. Non-homogeneous linear equations with constant coefficients

Let the equation

$$a_0y^{(n)} + a_1y^{(n-1)} + \dots + a_ny = f(x) \tag{7}$$

be given. The general solution of the non-homogeneous equation (7) is the sum of the general solution of the corresponding homogeneous equation and some particular solution of the non-homogeneous equation. Thus the problem of integrating equation (7) reduces to that of finding a particular solution, if the general solution of the homogeneous equation is known.

We note that in general the integration of equation (7) can be carried out by the method of variation of parameters. However, for functions $f(x)$ of a special type a particular solution may be found much more easily by inspection. The most general form of the right hand side $f(x)$ of equation (7) for which we may solve by inspection is the following:

$$f(x) = e^{\alpha x}[P_n(x) \cos \beta x + Q_m(x) \sin \beta x], \tag{8}$$

where $P_n(x)$ and $Q_m(x)$ are polynomials of degree n and m respectively. In this case we look for a particular solution* $y_{p.n.}$ of equation (7) of the form

$$y_{p.n.} = x^s e^{\alpha x}[\tilde{P}_k(x) \cos \beta x + \tilde{Q}_k(x) \sin \beta x], \tag{9}$$

where $k = \max(m, n)$, and $\tilde{P}_k(x)$ and $\tilde{Q}_k(x)$ are polynomials in x of degree k with undetermined coefficients, and s is the multiplicity of the roots $\lambda = \alpha \pm i\beta$ of the characteristic equation (if $\alpha \pm i\beta$ are not roots of the characteristic equation, then $s = 0$).

* The suffixes stand for particular, non-homogeneous.

Note. If the right hand side $f(x)$ consists of a sum

$$f(x) = \sum_{i=1}^{l} \alpha_i f_i(x), \tag{10}$$

where the $f_i(x)$ have the form (8), then by the principle of composition we look for a particular solution $y_{p.n.}$ of equation (7) of the form

$$y_{p.n.} = \sum_{i=1}^{l} \alpha_i y_{ip.n.}.$$

We sum up by means of a table of forms of the particular solution for various forms of the right hand side.

Right hand side of differential equation	Roots of characteristic equation	Form of particular solution
I. $P_m(x)$	1. The number 0 is not a root of the characteristic equation	$\tilde{P}_m(x)$
	2. The number 0 is a root of the characteristic equation of multiplicity s	$x^s\tilde{P}_m(x)$
II. $P_m(x)e^{\alpha x}$ (α real)	1. The number α is not a root of the characteristic equation	$\tilde{P}_m(x)e^{\alpha x}$
	2. The number α is a root of the characteristic equation of multiplicity s	$x^s\tilde{P}_m(x)e^{\alpha x}$
III. $P_n(x) \cos \beta x + Q_m(x) \sin \beta x$	1. The numbers $\pm i\beta$ are not roots of the characteristic equation	$\tilde{P}_k(x) \cos \beta x + \tilde{Q}_k(x) \sin \beta x$
	2. The numbers $\pm i\beta$ are roots of the characteristic equation of multiplicity s	$x^s(\tilde{P}_k(x) \cos \beta x + \tilde{Q}_k(x) \sin \beta x)$
IV. $e^{\alpha x}[P_n(x) \cos \beta x + Q_m(x) \sin \beta x]$	1. The numbers $\alpha \pm i\beta$ are not roots of the characteristic equation	$(\tilde{P}_k(x) \cos \beta x + \tilde{Q}_k(x) \sin \beta x)e^{\alpha x}$
	2. The numbers $\alpha \pm i\beta$ are roots of the characteristic equation of multiplicity s	$x^s(\tilde{P}_k(x) \cos \beta x + \tilde{Q}_k(x) \sin \beta x)e^{\alpha x}$

Example. Find the general solution of the equation

$$y''' - y'' + y' - y = x^2 + x.$$

Solution. The characteristic equation $\lambda^3 - \lambda^2 + \lambda - 1 = 0$ has distinct roots: $\lambda_1 = 1, \lambda_2 = -i, \lambda_3 = i$, so the general solution $y_{g.h.}$ of the corresponding homogeneous equation is

$$y_{g.h.} = C_1e^x + C_2 \cos x + C_3 \sin x.$$

Since the number 0 is not a root of the characteristic equation, we look for a particular solution $y_{p.n.}$ of the given equation of the form

$$y_{p.n.} = A_1x^2 + A_2x + A_3,$$

where A_1, A_2, A_3 are unknown coefficients, to be determined. Substituting the expression for $y_{p.n.}$ into the given equation, we obtain

$$-A_1x^2 + (2A_1 - A_2)x + (A_2 - 2A_1 - A_3) = x^2 + x,$$

hence

$$\left.\begin{aligned} A_1 &= -1 \\ 2A_1 - A_2 &= 1 \\ A_2 - 2A_1 - A_3 &= 0 \end{aligned}\right\}.$$

Solving this system, we find that

$$A_1 = -1, \quad A_2 = -3, \quad A_3 = -1,$$

and so a particular solution is

$$y_{p.n.} = -x^2 - 3x - 1$$

and the general solution $y_{g.n.}$ of the given equation has the form

$$y_{g.n.} = C_1e^x + C_2 \cos x + C_3 \sin x - x^2 - 3x - 1.$$

Example. Find the general solution of the equation

$$y''' - y'' = 12x^2 + 6x.$$

Solution. The characteristic equation $\lambda^3 - \lambda^2 = 0$ has roots $\lambda_1 = \lambda_2 = 0, \lambda_3 = 1$, and so the general solution of the corresponding homogeneous equation is

$$y_{g.h.} = C_1 + C_2x + C_3e^x.$$

Since the number 0 is a double root of the characteristic equation, we must look for a particular solution of the form

$$y_{p.n.} = x^2(A_1x^2 + A_2x + A_3) = A_1x^4 + A_2x^3 + A_3x^2.$$

Substituting the expression for $y_{p.n.}$ into the given equation, we have

$$-12A_1x^2 + (24A_1 - 6A_2)x + (6A_2 - 2A_3) = 12x^2 + 6x,$$

$$\left.\begin{aligned} -12A_1 &= 12 \\ 24A_1 - 6A_2 &= 6 \\ 6A_2 - 2A_3 &= 0 \end{aligned}\right\} .$$

This system has the solution: $A_1 = -1$, $A_2 = -5$, $A_3 = -15$, and so

$$y_{p.n.} = -x^4 - 5x^3 - 15x^2.$$

The general solution of the given equation is

$$y_{g.n.} = C_1 + C_2x + C_3e^x - x^4 - 5x^3 - 15x^2.$$

Example. $y'' - 6y' + 9y = 25e^x \sin x$.

Solution. The characteristic equation $\lambda^2 - 6\lambda + 9 = 0$ has roots $\lambda_1 = \lambda_2 = 3$; the general solution $y_{g.h.}$ of the homogeneous equation is $y_{g.h.} = (C_1 + C_2x)e^{3x}$. We look for a particular solution $y_{p.n.}$ of the form

$$y_{p.n.} = e^x(a \cos x + b \sin x).$$

Substituting the expression for $y_{p.n.}$ into the equation, and cancelling e^x from both sides of the equation, we obtain

$$(3a - 4b) \cos x + (4a + 3b) \sin x = 25 \sin x.$$

Hence we have

$$\left.\begin{aligned} 3a - 4b &= 0 \\ 4a + 3b &= 25 \end{aligned}\right\} .$$

The solution of this system is: $a = 4$, $b = 3$, and so

$$y_{p.n.} = e^x(4 \cos x + 3 \sin x).$$

The general solution of the given equation is

$$y_{g.n.} = (C_1 + C_2x)e^{3x} + e^x(4 \cos x + 3 \sin x).$$

In the case where the right hand side $f(x)$ contains trigonometric functions $\sin \beta x$ and $\cos \beta x$ it is convenient to go over to exponential functions: we demonstrate this by means of an example.

Suppose we require the solution of the differential equation

$$y'' + y = x \cos x. \tag{1}$$

Here $\lambda^2 + 1 = 0$, $\lambda_1 = -i$, $\lambda_2 = i$, and the general solution of the homogeneous equation has the form

$$y_{g.h.} = C_1 \cos x + C_2 \sin x.$$

We must look for a particular solution $y_{p.n.}$ of the non-homogeneous equation in the form

$$y_{p.n.} = x[(A_1x + A_2) \cos x + (B_1x + B_2) \sin x].$$

We proceed like this. We consider the equation

$$z'' + z = xe^{ix}. \tag{2}$$

It is easy to see that the right hand side of the original equation is the real part of the right hand side of equation (2) (Re xe^{ix}). We shall use the fact that if the differential equation with real coefficients

$$L[y] = f_1(x) + if_2(x)$$

has a solution $y = u(x) + iv(x)$, then $u(x)$ is a solution of the equation $L[y] = f_1(x)$ and $v(x)$ is a solution of the equation $L[y] = f_2(x)$.

We find $z_{p.n.}$ from equation (2):

$$z_{p.n.} = (Ax + B)xe^{ix} = (Ax^2 + Bx)e^{ix}$$

$$z''_{p.n.} = 2Ae^{ix} + 2(2Ax + B)ie^{ix} - (Ax^2 + Bx)e^{ix}.$$

Substituting into equation (2) and cancelling e^{ix} from both sides, we have

$$2A + 4Axi + 2Bi = x,$$

hence

$$4Ai = 1, \quad A = -\frac{i}{4},$$

$$A + Bi = 0, \quad B = -\frac{A}{i} = \frac{1}{4}.$$

So

$$z_{p.n.} = \left(-\frac{i}{4}x^2 + \frac{1}{4}x\right) e^{ix} = \left(-\frac{i}{4}x^2 + \frac{1}{4}x\right) (\cos x + i \sin x)$$

$$= \frac{x \cos x + x^2 \sin x}{4} + i\frac{x \sin x - x^2 \cos x}{4},$$

$$\text{Re } z_{p.n.} = \frac{x \cos x + x^2 \sin x}{4} = y_{p.n.} \text{ of equation (1).}$$

This method sometimes considerably simplifies and shortens the calculation associated with finding particular solutions.

Example.

$$y'' + 4y = \sin 2x. \tag{1}$$

We consider the equations

$$z'' + 4z = e^{2ix}, \tag{2}$$

$$\sin 2x = \operatorname{Im} e^{2ix},$$

$$z_{p.n.} = Axe^{2ix},$$

$$z''_{p.n.} = -4Axe^{2ix} + 4iAe^{2ix}.$$

Substituting into equation (1), we obtain

$$4iA = 1, \quad A = -\frac{i}{4},$$

$$z_{p.n.} = -\frac{i}{4}xe^{2ix} = -\frac{i}{4}x(\cos 2x + i\sin 2x)$$

$$= \frac{1}{4}x\sin 2x - i\frac{x}{4}\cos 2x.$$

$$\operatorname{Im} z_{p.n.} = -\frac{x}{4}\cos 2x = y_{p.n.} \text{ of equation (1).}$$

Determine the form of a particular solution of the non-homogeneous linear equation, given the roots of the characteristic equation and the right hand side $f(x)$:

468. $\lambda_1 = 1, \lambda_2 = 2;\ f(x) = ax^2 + bx + c.$

469. $\lambda_1 = 0, \lambda_2 = 1;\ f(x) = ax^2 + bx + c.$

470. $\lambda_1 = 0, \lambda_2 = 0;\ f(x) = ax^2 + bx + c.$

471. $\lambda_1 = 1, \lambda_2 = 2;\ f(x) = e^{-x}(ax + b).$

472. $\lambda_1 = -1, \lambda_2 = 1;\ f(x) = e^{-x}(ax + b).$

473. $\lambda_1 = -1, \lambda_2 = -1;\ f(x) = e^{-x}(ax + b).$

474. $\lambda_1 = 0, \lambda_2 = 1;\ f(x) = \sin x + \cos x.$

475. $\lambda_1 = -i, \lambda_2 = i;\ f(x) = \sin x + \cos x.$

476. $\lambda_1 = -2i, \lambda_2 = 2i;\ f(x) = a\sin 2x + b\cos 2x.$

477. $\lambda_1 = -ki, \lambda_2 = ki;\ f(x) = a\sin kx + b\cos kx.$

478. $\lambda_1 = 1, \lambda_2 = 1;\quad f(x) = e^{-x}(a \sin x + b \cos x).$

479. $\lambda_1 = -1 - i, \lambda_2 = -1 + i;\quad f(x) = e^{-x}(a \sin x + b \cos x).$

480. $\lambda_1 = \lambda_2 = \lambda_3 = 1;\quad f(x) = ax^2 + bx + c.$

481. $\lambda_1 = 0, \lambda_2 = 1, \lambda_3 = 2;\quad f(x) = ax^2 + bx + c.$

482. $\lambda_1 = \lambda_2 = 0, \lambda_3 = 1;\quad f(x) = ax^2 + bx + c.$

483. $\lambda_1 = \lambda_2 = \lambda_3 = 0;\quad f(x) = ax^2 + bx + c.$

484. $\lambda_1 = i, \lambda_2 = -i, \lambda_3 = 1;\quad f(x) = \sin x + \cos x.$

485.
$$\left.\begin{array}{ll} (a) & \lambda_1 = -1, \lambda_2 = 1, \lambda_3 = 2 \\ (b) & \lambda_1 = 0, \lambda_2 = 2, \lambda_3 = 3 \\ (c) & \lambda_1 = \lambda_2 = -1, \lambda_3 = 1 \\ (d) & \lambda_1 = \lambda_2 = \lambda_3 = -1 \\ (e) & \lambda_1 = -1, \lambda_2 = \lambda_3 = 1 \\ (f) & \lambda_1 = \lambda_2 = \lambda_3 = 1 \end{array}\right\} f(x) = ae^{-x} + be^{x}.$$

486.
$$\left.\begin{array}{ll} (a) & \lambda_1 = 0, \lambda_2 = 1 \\ (b) & \lambda_1 = k, \lambda_2 = 1 \\ (c) & \lambda_1 = \lambda_2 = k \\ (d) & \lambda_1 = \lambda_2 = 0\ \ \lambda_3 = 1 \\ (e) & \lambda_1 = \lambda_2 = k, \lambda_3 = 1 \\ (f) & \lambda_1 = \lambda_2 = \lambda_3 = k \end{array}\right\} f(x) = (ax^2 + bx + c)e^{kx}.$$

487.
$$\left.\begin{array}{ll} (a) & \lambda_1 = \lambda_2 = 1, \lambda_3 = 2 \\ (b) & \lambda_1 = -i, \lambda_2 = i, \lambda_3 = 0 \end{array}\right\} f(x) = a \sin x + b \cos x.$$

488.
$$\left.\begin{array}{ll} (a) & \lambda_1 = 3 - 2i, \lambda_2 = 3 + 2i \\ & \lambda_3 = \lambda_4 = 0 \\ (b) & \lambda_1 = \lambda_2 = 3 - 2i \\ & \lambda_3 = \lambda_4 = 3 + 2i \end{array}\right\} f(x) = e^{3x}(\sin 2x + \cos 2x).$$

For the following non-homogeneous linear equations, determine the form of a particular solution:

489. $y'' + 3y' = 3.$

490. $y'' - 7y' = (x - 1)^2.$

491. $y'' + 3y' = e^{x}.$

492. $y'' + 7y' = e^{-7x}$.

493. $y'' - 8y' + 16y = (1 - x)e^{4x}$.

494. $y'' - 10y' + 25y = e^{5x}$.

495. $4y'' - 3y' = xe^{3x/4}$.

496. $y'' - y' - 2y = e^{x} + e^{-2x}$.

497. $y'' - 4y' = xe^{4x}$.

498. $y'' + 25y = \cos 5x$.

499. $y'' + y = \sin x - \cos x$.

500. $y'' + 16y = \sin(4x + \alpha)$.

501. $y'' + 4y' + 8y = e^{2x}(\sin 2x + \cos 2x)$.

502. $y'' - 4y' + 8y = e^{2x}(\sin 2x - \cos 2x)$.

503. $y'' + 6y' + 13y = e^{-3x} \cos 2x$.

504. $y'' + k^2 y = k \sin(kx + \alpha)$.

505. $y'' + k^2 y = k$.

506. $y'' + 4y = \sin x \sin 2x$.

507. $y'' - 4y' = 2 \cos^2 4x$.

508. $y''' + y = x$.

509. $y''' + 6y'' + 11y' + 6y = 1$.

510. $y''' + y' = 2$.

511. $y''' + y'' = 3$.

512. $y^{IV} - y = 1$.

513. $y^{IV} - y' = 2$.

514. $y^{IV} - y'' = 3$.

515. $y^{IV} - y''' = 4$.

516. $y^{IV} + 4y''' + 4y'' = 1$.

517. $y^{IV} + 2y''' + y'' = e^{4x}$.

518. $y^{IV} + 2y''' + y'' = e^{-x}$.

519. $y^{IV} + 2y''' + y'' = xe^{-x}$.

520. $y^{IV} + 4y'' + 4y = \sin 2x$.

521. $y^{IV} + 4y'' + 4y = \cos x$.

522. $y^{IV} + 4y'' + 4y = x \sin 2x$.

523. $y^{IV} + 2n^2y'' + n^4y = a \sin (nx + \alpha)$.

524. $y^{IV} - 2n^2y'' + n^4y = \cos (nx + \alpha)$.

525. $y^{IV} + 4y''' + 6y'' + 4y' + y = \sin x$.

526. $y^{IV} - 4y''' + 6y'' - 4y' + y = e^x$.

527. $y^{IV} - 4y''' + 6y'' - 4y' + y = xe^x$.

Solve the following equations:

528. $y'' + 2y' + y = -2$.

529. $y'' + 2y' + 2 = 0$.

530. $y'' + 9y - 9 = 0$.

531. $y''' + y'' = 1$.

532. $5y''' - 7y'' - 3 = 0$.

533. $y^{IV} - 6y''' + 6 = 0$.

534. $3y^{IV} + y''' = 2$.

535. $y^{IV} - 2y''' + 2y'' - 2y' + y = 1$.

536. $y'' - 4y' + 4y = x^2$.

537. $y'' + 8y' = 8x$.

538. $y'' - 2ky' + k^2y = e^x$.

539. $y'' + 4y' + 4y = 8e^{-2x}$.

540. $y'' + 4y' + 3y = 9e^{-3x}$.

541. $7y'' - y' = 14x$.

542. $y'' + 3y' = 3xe^{-3x}$.

543. $y'' + 5y' + 6y = 10(1 - x)e^{-2x}$.

544. $y'' + 2y' + 2y = 1 + x$.

545. $y'' + y' + y = (x + x^2)e^x$.

546. $y'' + 4y' - 2y = 8 \sin 2x$.

547. $y'' + y = 4x \cos x$.

548. $y'' - 2my' + m^2y = \sin nx$.

549. $y'' + 2y' + 5y = e^{-x} \sin 2x$.

550. $y'' + a^2y = 2 \cos mx + 3 \sin mx \qquad (m \neq a)$.

551. $y'' - y' = e^x \sin x$.

552. $y'' + 2y' = 4e^x(\sin x + \cos x)$.

553. $y'' + 4y' + 5y = 10e^{-2x} \cos x$.

554. $y'' + 2y' + 5y = e^{-x}(2x + \sin 2x)$.

555. $4y'' + 8y' = x \sin x$.

556. $y'' - 3y' + 2y = xe^x$.

557. $y'' + y' - 2y = x^2e^{4x}$.

558. $y'' - 3y' + 2y = (x^2 + x)e^{3x}$.

559. $y''' - y'' + y' - y = x^2 + x$.

560. $y^{\mathrm{IV}} - 2y''' + 2y'' - 2y' + y = e^x$.

561. $y'' - 2y' + y = x^3$.

562. $5y'' - 6y' + 5y = 13e^x \operatorname{ch} x$.

563. $y^{\mathrm{IV}} + y'' = x^2 + x$.

564. $y^{\mathrm{V}} - y^{\mathrm{IV}} = xe^x - 1$.

565. $y'' + y = x^2 \sin x$.

566. $y'' + 2y' + y = x^2e^{-x} \cos x$.

567. $y''' - 4y' = xe^{2x} + \sin x + x^2$.

568. $y''' - y = \sin x$.

569. $y'' + 2y' + 2y = e^{-x} \cos x + xe^{-x}$.

570. $y^{\mathrm{IV}} - 2y'' + y = \cos x$.

571. $y'' + y = 2 \sin x \sin 2x$.

572. $y'' + 4y = x \sin^2 x$.

573. $y^{\mathrm{IV}} + 2y''' + 2y'' + 2y' + y = xe^x + \frac{1}{2} \cos x$.

574. $y'' + y' = \cos^2 x + e^x + x^2$.

575. $y^{\mathrm{V}} + 4y''' = e^x + 3\sin 2x + 1.$

576. $y''' - 3y'' + 3y' - y = e^x \cos 2x.$

577. $y''' - 2y' + 4y = e^x \cos x + x^2 + \sin 2x.$

In the following problems, find particular solutions of the equations satisfying the given initial conditions:

578. $y'' - 5y' + 6y = (12x - 7)e^{-x}; \quad y(0) = y'(0) = 0.$

579. $y'' + 9y = 6e^{3x}; \quad y(0) = y'(0) = 0.$

580. $y'' - 4y' + 5y = 2x^2e^x; \quad y(0) = 2,\ y'(0) = 3.$

581. $y'' + 6y' + 9y = 10\sin x; \quad y(0) = y'(0) = 0.$

582. $y'' + y = 2\cos x; \quad y(0) = 1,\ y'(0) = 0.$

583. $y'' + 4y = \sin x; \quad y(0) = y'(0) = 1.$

584. $y'' - 6y' + 9y = x^2 - x + 3; \quad y(0) = \frac{4}{3},\ y'(0) = \frac{1}{27}.$

585. $y'' - 4y' + 4y = e^{2x}; \quad y(0) = 2,\ y'(0) = 8.$

586. $y'' + 4y = 4(\sin 2x + \cos 2x); \quad y(\pi) = y'(\pi) = 2\pi.$

587. $y'' - y' = -5e^{-x}(\sin x + \cos x); \quad y(0) = -2,\ y'(0) = 3.$

588. $y'' - 2y' + 2y = 2e^x(\sin x + \cos x); \quad y(\pi) = \pi e^\pi,\ y'(\pi) = e^\pi.$

589. $y''' - y' = -2x; \quad y(0) = 0,\ y'(0) = 2,\ y''(0) = 2.$

590. $y^{\mathrm{IV}} - y = 8e^x;$
$y(0) = -1,\ y'(0) = 0,\ y''(0) = 1,\ y'''(0) = 0.$

591. $y''' + y' = 2x; \quad y(0) = y'(0) = 0,\ y''(0) = 2.$

592. $y^{\mathrm{IV}} - y = 8e^x; \quad y(0) = 0,\ y'(0) = 2,\ y''(0) = 4,\ y'''(0) = 6.$

In the following problems, find particular solutions of the equations satisfying the given conditions at infinity:

593. $y'' - 4y' + y = \sin x, \quad y$ bounded as $x \to +\infty.$

594. $y'' + 2y' + 5y = 4\cos 2x + \sin 2x,$
y bounded as $x \to -\infty.$

595. $y'' - y = 1, \quad y$ bounded as $x \to \infty.$

596. $y'' - y = -2\cos x, \quad y$ bounded as $x \to \infty.$

597. $y'' - 2y' + y = 4e^{-x}, \quad y \to 0$ as $x \to +\infty.$

598. $y'' + 4y' + 3y = 8e^x + 9, \quad y \to 3 \quad \text{as} \quad x \to -\infty.$

599. $y'' - y' - 5y = 1, \quad y \to -\frac{1}{5} \quad \text{as} \quad x \to \infty.$

600. $y'' + 4y' + 4y = 2e^x(\sin x + 7\cos x),$
$y \to 0 \quad \text{as} \quad x \to -\infty.$

601. $y'' - 5y' + 6y = 2e^{-2x}(9\sin 2x + 8\cos 2x),$
$y \to 0 \quad \text{as} \quad x \to +\infty.$

602. $y'' - 4y' + 4y = (9x^2 - 3x - 4)e^{-x},$
$y \to 0 \quad \text{as} \quad x \to +\infty.$

4. Euler equations

An equation of the form

$$a_0x^ny^{(n)} + a_1x^{n-1}y^{(n-1)} + \ldots + a_{n-1}xy' + a_ny = 0, \tag{12}$$

where all the a_i are constants, is called an Euler equation.

By a change of variable

$$x = e^t$$

this equation is transformed into a homogeneous linear equation with constant coefficients:

$$b_0y_t^{(n)} + b_1y_t^{(n-1)} + \ldots + b_{n-1}y_t' + b_ny(t) = 0. \tag{13}$$

Note 1. An equation of the form

$$a_0(ax+b)^ny^{(n)} + a_1(ax+b)^{n-1}y^{(n-1)} + \ldots + a_{n-1}(ax+b)y' + a_ny = 0 \tag{14}$$

is also called an Euler equation and is reduced to a homogeneous linear equation with constant coefficients by means of a change of variables

$$ax + b = e^t.$$

Note 2. We can look for a particular solution of equation (12) of the form

$$y = x^k,$$

where for k we obtain an equation which coincides with the characteristic equation of equation (13).

We give now some examples of solutions of Euler equations.

Example. Find the general solution of the Euler equation

$$x^2y'' + 2xy' - 6y = 0.$$

First method. We make the substitution $x = e^t$; then

$$y' = \frac{dy}{dx} = \frac{\frac{dy}{dt}}{\frac{dx}{dt}} = e^{-t}\frac{dy}{dt},$$

$$y'' = \frac{dy'}{dx} = \frac{\frac{dy'}{dt}}{\frac{dx}{dt}} = \frac{\left(\frac{d^2y}{dt^2} - \frac{dy}{dt}\right)e^{-t}}{e^t} = e^{-2t}\left(\frac{d^2y}{dt^2} - \frac{dy}{dt}\right)$$

and the equation takes the form

$$\frac{d^2y}{dt^2} + \frac{dy}{dt} - 6y = 0.$$

The roots of the characteristic equation are: $\lambda_1 = -3$, $\lambda_2 = 2$, and the general solution of the last equation is: $y = C_1e^{-3t} + C_2e^{2t}$. But since $x = e^t$, we have $y = C_1x^{-3} + C_2x^2$ or

$$y = \frac{C_1}{x^3} + C_2x^2.$$

Second method. We look for a solution of the given equation of the form $y = x^k$, where k is unknown. We find that

$$y' = kx^{k-1}, \quad y'' = k(k-1)x^{k-2}.$$

Substituting into the equation, we obtain

$$x^2k(k-1)x^{k-2} + 2xkx^{k-1} - 6x^k = 0$$

or

$$x^k[k(k-1) + 2k - 6] = 0.$$

But since $x^k \not\equiv 0$, we have $k(k-1) + 2k - 6 = 0$ or $k^2 + k - 6=0$. The roots of this equation are $k_1 = -3$, $k_2 = 2$. The corresponding fundamental system of solutions is:

$$y_1 = x^{-3}, \quad y_2 = x^2,$$

and the general solution has the form

$$y = C_1x^{-3} + C_2x^2.$$

Integrate the following Euler equations:

603. $x^2y'' + xy' - y = 0$.

604. $x^2y'' + 3xy' + y = 0$.

605. $x^2y'' + 2xy' + 6y = 0$.

606. $xy'' + y' = 0$.

607. $(x - 2)^2y'' + 3(x - 2)y' - 3y = 0$.

608. $(2x + 1)^2y'' - 2(2x + 1)y' + 4y = 0$.

609. $x^2y''' - 3xy'' + 3y' = 0$.

610. $x^2y''' = 2y'$.

611. $(x + 1)^2y''' - 12y' = 0$.

612. $(2x + 1)^2y''' + 2(2x + 1)y'' + 2y' = 0$.

5. Linear differential equations with variable coefficients

If a particular solution $y_1(x)$ of the equation

$$y^{(n)} + p_1(x)y^{(n-1)} + \dots + p_n(x)y = 0, \tag{1}$$

is known, the order of the equation can be reduced by one (without affecting the linearity of the equation) by putting $y = y_1z$, where z is a new unknown function, and then making the substitution $z' = u$ [we could straight away make the substitution $u = (y/y_1)'$].

If k linearly independent particular solutions of equation (1) are known, the order of the equation can be reduced by k.

For certain linear equations with rational coefficients it is advisable to look for particular solutions in the form of polynomials with undetermined coefficients.

For an equation of the form

$$a(x)y'' + b(x)y' + c(x)y = 0,$$

where $a(x)$, $b(x)$, $c(x)$ are polynomials, the general integral is a fraction. The denominator of such an integral may consist only of linear factors, which are divisors of the polynomial $a(x)$.

The general solution of the equation

$$y^{(n)} + p_1(x)y^{(n-1)} + \dots + p_n(x)y = f(x) \tag{2}$$

is the sum of any particular solution and the general solution of the corresponding homogeneous equation (1).

If the fundamental system of the corresponding homogeneous

equation (1) is known, the general solution of the non-homogeneous equation (2) can be found by means of quadratures by the method of variation of parameters.

The general solution of equation (1) has the form

$$y = C_1y_1 + C_2y_2 + \dots + C_ny_n, \tag{3}$$

where $C_1, C_2, \dots, C_n$ are arbitrary constants.

We look for a solution of (2) in the form

$$y = C_1(x)y_1 + C_2(x)y_2 + \dots + C_n(x)y_n, \tag{4}$$

where $C_1(x), C_2(x), \dots, C_n(x)$ are certain undetermined functions of x.

To determine them, we obtain the following system

$$\left.\begin{array}{l} y_1C_1' + y_2C_2' + \dots + y_nC_n' = 0 \\ y_1'C_1' + y_2'C_2' + \dots + y_n'C_n' = 0 \\ \cdot\quad\cdot\quad\cdot\quad\cdot\quad\cdot\quad\cdot\quad\cdot\quad\cdot\quad\cdot \\ y_1^{(n-1)}C_1' + y_2^{(n-1)}C_2' + \dots + y_n^{(n-1)}C_n' = f(x) \end{array}\right\}. \tag{5}$$

Solving this system for C_i' $(i = 1, 2, \dots, n)$, we obtain

$$\frac{dC_i}{dx} = \phi_i(x) \qquad (i = 1, 2, \dots, n).$$

Hence

$$C_i(x) = \int\phi_i(x)\,dx + \tilde{C}_i \qquad (i = 1, 2, \dots, n). \tag{6}$$

Putting this value for $C_i(x)$ into (4), we obtain the general solution of equation (2).

For an equation of the second order

$$y'' + p_1(x)y' + p_2(x)y = f(x)$$

the system (5) has the form

$$\left.\begin{array}{l} y_1C_1' + y_2C_2' = 0, \\ y_1'C_1' + y_2'C_2' = f(x) \end{array}\right\}. \tag{5'}$$

Solving (5′) for C_1' and C_2', we obtain

$$C_1' = -\frac{y_2f(x)}{W[y_1, y_2]}, \qquad C_2' = \frac{y_1f(x)}{W[y_1, y_2]},$$

and so we find that

$$C_1(x) = -\int \frac{y_2 f(x)}{W[y_1, y_2]}\,dx + \tilde{C}_1, \qquad C_2(x) = \int \frac{y_1 f(x)}{W[y_1, y_2]}\,dx + \tilde{C}_2,$$

where $\tilde{C}_1$ and $\tilde{C}_2$ are constants of integration.

Example. Find the general solution of the equation

$$xy'' + 2y' + xy = 0, \text{ if } y_1 = \frac{\sin x}{x} \text{ is a particular solution.}$$

Solution. We put $y = \dfrac{\sin x}{x} \,.\, z$, where z is a new unknown function of x. We have

$$y' = y_1'z + y_1z', \quad y'' = y_1''z + 2y_1'z' + y_1z''.$$

Substituting into the given equation, we obtain

$$(xy_1'' + 2y_1' + xy_1)z + xy_1z'' + 2(xy_1' + y_1)z' = 0.$$

But since $y_1 = \dfrac{\sin x}{x}$ is a particular solution of the given equation, $xy_1'' + 2y_1' + xy_1 = 0$, and so we have

$$xy_1z'' + 2(xy_1' + y_1)z' = 0. \tag{1}$$

But $y_1' = \dfrac{\cos x}{x} - \dfrac{\sin x}{x^2}$, and so $xy_1' + y_1 = \cos x$, and equation (1) takes the form

$$z'' \sin x + 2z' \cos x = 0.$$

We write the equation in the form

$$\frac{z''}{z'} + 2\frac{\cos x}{\sin x} = 0.$$

Hence we have

$$(\ln |z'| + 2 \ln |\sin x|)' = 0,$$

and so

$$\ln |z'| + 2 \ln |\sin x| = \ln \tilde{C}_1 \quad \text{or} \quad z' \sin^2 x = \tilde{C}_1.$$

Integrating this equation, we find that

$$z = -\tilde{C}_1 \cot x + C_2,$$

and consequently the general solution of the given equation is

$$y = -\tilde{C}_1 \frac{\cos x}{x} + C_2 \frac{\sin x}{x}$$

or

$$y = C_1 \frac{\cos x}{x} + C_2 \frac{\sin x}{x} \qquad (C_1 = -\tilde{C}_1).$$

Example. Find the general solution of the equation

$$y'' + \frac{2}{x} y' + y = \frac{1}{x}.$$

Solution. The general solution of the corresponding homogeneous equation has the form (see example 1)

$$y_{g.h.} = C_1 \frac{\sin x}{x} + C_2 \frac{\cos x}{x}$$

and consequently its fundamental system of solutions is

$$y_1 = \frac{\sin x}{x}, \qquad y_2 = \frac{\cos x}{x}.$$

We shall look for the general solution of the given equation by the method of variation of parameters:

$$y = C_1(x) \frac{\sin x}{x} + C_2(x) \frac{\cos x}{x},$$

where $C_1(x)$, $C_2(x)$ are unknown functions of x, to be determined. To find them, we form the following system:

$$\left.\begin{aligned} &C_1'(x) \frac{\sin x}{x} + C_2'(x) \frac{\cos x}{x} = 0. \\ &C_1'(x) \frac{x \cos x - \sin x}{x^2} + C_2'(x) \frac{-x \sin x - \cos x}{x^2} = \frac{1}{x} \end{aligned}\right\}.$$

Hence we find that

$$C_1'(x) = \cos x, \quad C_2'(x) = -\sin x.$$

Integrating, we obtain

$$C_1(x) = \sin x + \tilde{C}_1, \quad C_2(x) = \cos x + \tilde{C}_2.$$

Substituting these values for $C_1(x)$ and $C_2(x)$ into the expression for y, we find the general integral of the given equation:

$$y = \tilde{C}_1 \frac{\sin x}{x} + \tilde{C}_2 \frac{\cos x}{x} + \frac{1}{x}.$$

Example. Solve the equation

$$y'' + y = \frac{1}{\cos x}.$$

Solution. The corresponding homogeneous equation is $y'' + y = 0$. Its characteristic equation has the imaginary roots $\lambda_1 = -i$, $\lambda_2 = i$, and the general solution of the homogeneous equation has the form:

$$y_{g.h.} = C_1 \cos x + C_2 \sin x.$$

We look for the general solution of the original equation in the form

$$y = C_1(x) \cos x + C_2(x) \sin x, \qquad (A)$$

where $C_1(x)$ and $C_2(x)$ are unknown functions of x. To find them, we form the system

$$\left.\begin{aligned} \cos x \, . \, C_1'(x) + \sin x \, . \, C_2'(x) &= 0 \\ -\sin x \, . \, C_1'(x) + \cos x \, . \, C_2'(x) &= \frac{1}{\cos x} \end{aligned}\right\}.$$

We solve this system for $C_1'(x)$ and $C_2'(x)$:

$$C_1'(x) = -\tan x; \qquad C_2'(x) = 1.$$

Hence, by integration, we obtain

$$C_1(x) = \ln |\cos x| + \tilde{C}_1, \qquad C_2(x) = x + \tilde{C}_2.$$

Substituting the expressions for $C_1(x)$ and $C_2(x)$ into (A), we obtain the general integral of the given equation

$$y = \tilde{C}_1 \cos x + \tilde{C}_2 \sin x + \cos x \, . \ln |\cos x| + x \sin x.$$

Here $\cos x \ln |\cos x| + x \sin x$ is a particular solution of the original non-homogeneous equation.

Integrate the following equations (y_1, y_2 are particular solutions of the homogeneous equation):

613. $x^3 y''' - 3x^2 y'' + 6xy' - 6y = 0$; $y_1 = x$, $y_2 = x^2$.

614. $(x^2 - 1)y'' = 6y$; y_1 is a polynomial.

615. $(2x + 1)y'' + (4x - 2)y' - 8y = 0$; $y_1 = e^{mx}$.

616. $(x^2 - x)y'' + (2x - 3)y' - 2y = 0$; y_1 is a rational fraction whose denominator is a power of a divisor of the coefficient of y''.

617. $(3x + 2x^2)y'' - 6(1 + x)y' + 6y = 6.$

618. $x^2(\ln x - 1)y'' - xy' + y = 0, \quad y_1 = x.$

619. $y'' - 4xy' + (4x^2 - 3)y = 0.$

620. $y'' + y' \tan x + y \cos^2 x = 0.$

621. $(1 + x^2)y'' + xy' - y + 1 = 0; \quad y_1 = x.$

622. $x^2y'' - xy' - 3y = 5x^4, \quad y_1 = \dfrac{1}{x}.$

623. $(4x^2 - x)y'' + 2(2x - 1)y' - 4y = 12x^2 - 6x, \quad y_1 = \dfrac{1}{x}.$

624. $y'' - y' + ye^{2x} = xe^{2x} - 1, \quad y_1 = \sin e^x.$

625. $y'' + y' \tan x = \dfrac{\cos^2 x}{\sin x}.$

626. $(x + 1)^3y'' + 3(x + 1)^2y' + (x + 1)y = 6 \ln (x + 1).$

627. $x(x - 1)y'' - (2x - 1)y' + 2y = x^2(2x - 3), \quad y_1 = x^2.$

628. A chain of length 6 m falls from a smooth table. If the motion begins at the instant when 1m. of chain hangs down, how long will it take for the whole chain to fall off?

629. Find the equation of motion of a particle, if the acceleration is given in terms of the time by the formula $a = 1{\cdot}2\, t$, and if the distance s satisfies the conditions: $s = 0$ when $t = 0$, $s = 20$ when $t - 5$.

630. A body of mass m slides on a horizontal plane, having been acted on by an impulse which gives it an initial speed v_0. A frictional force equal to km acts on the body. Find the distance which the body travels.

631. A particle of mass $m = 1$ moves along a straight line towards a centre which repels it with a force equal to k^2x (x is the distance of the particle from the centre). When $t = 0$, $x = a$, $dx/dt = -ka$. Find the law of motion.

Integrate the following equations by the method of variation of parameters:

632. $y'' + 4y = \dfrac{1}{\cos 2x}.$

633. $y'' + y = \tan^2 x.$

634. $y'' - y = \dfrac{2e^x}{e^x - 1}.$

635. $y'' - y' = \dfrac{1}{e^x + 1}.$

636. $y'' + y = \dfrac{1}{\sqrt{\sin^5 x \cos x}}.$

637. $y'' + y = \dfrac{1}{(\cos 2x)^{3/2}}.$

638. $y''' - 2y'' - y' + 2y = \dfrac{2x^3 + x^2 - 4x - 6}{x^4}.$

639. $y'' + y = \dfrac{1}{\sqrt[3]{\sin^7 x \cos^8 x}}.$

640. $y'' - 2y' + y = \dfrac{e^x}{x^2 + 1}.$

641. $y'' + 2y' + 2y = \dfrac{1}{e^x \sin x}.$

642. $y'' - y' = e^{2x} \cos e^x.$

643. $y'' + y' = -\dfrac{1}{x}.$

644. $y'' + 3y' + 2y = \dfrac{2x}{(x + 1)^2}.$

645. $y'' + y = \dfrac{1}{x^2}.$

646. $xy'' - (1 + 2x^2)y' = 4x^3 e^{x^2}.$

647. $y'' - 2y' \tan x = 1.$

648. $y'' x \ln x - y' = \ln^2 x.$

649. $xy'' + (2x - 1)y' = -4x^2.$

650. $(x - 1)y'' - xy' + y = (x - 1)^2 e^x; \quad y_1 = e^x.$

651. $y'' + y' + e^{-2x} y = e^{-3x}; \quad y_1 = \cos e^{-x}.$

652. $(x^4 - x^3)y'' + (2x^3 - 2x^2 - x)y' - y = \dfrac{(x - 1)^2}{x}; \quad y_1 = \dfrac{1}{x}.$

In the following problems the fundamental system y_1, y_2 of the corresponding homogeneous equation is given.

653. $y''(\cos x - \sin x) + 2y' \sin x - y(\cos x + \sin x)$
$$= e^x(\cos x - \sin x)^2; \quad y_1 = e^x,\ y_2 = \sin x.$$

654. $xy'' - y' - 4x^3y = 16x^3e^{x^2}; \quad y_1 = e^{x^2},\ y_2 = e^{-x^2}.$

655. $x(1 - x\ln x)y'' + (1 + x^2 \ln x)y' - (x + 1)y = (1 - x\ln x)^2e^x;$
$y_1 = e^x,\ y_2 = \ln x.$

656. $4(x^2 + x)y'' + 2(2x + 1)y' - y = 2\sqrt{x^2 + x},$
$$y|_{x=1} = \frac{2\sqrt{2}}{3}, \quad y'|_{x=1} = \frac{1}{\sqrt{2}}; \quad y_1 = \sqrt{x},\ y_2 = \sqrt{x + 1}.$$

657. $y'' \cos^2 x - y' \sin x \cos x - y = \sin x, \quad y|_{x=0} = y'|_{x=0} = 1;$
$y_1 = \sec x,\ y_2 = \tan x.$

658. $y'' \sin x + 2y' \cos x - y \sin x = 2\cos 2x,$
$$y|_{x=\frac{\pi}{2}} = \tfrac{1}{2}, \quad y'|_{x=\frac{\pi}{2}} = 0; \quad y_1 = \frac{x}{\sin x}, \quad y_2 = \frac{1}{\sin x}.$$

659. $4xy'' + 2y' + y = 1, \quad \lim\limits_{x\to+\infty} y = 1;$
$y_1 = \sin\sqrt{x},\ y_2 = \cos\sqrt{x}.$

660. $4xy'' + 2y' + y = \dfrac{6 + x}{x^2}, \quad \lim\limits_{x\to+\infty} y = 0.$

661. $(1 + x^2)y'' + 2xy' = \dfrac{1}{1 + x^2}, \quad \lim\limits_{x\to+\infty} y = \dfrac{\pi^2}{8}, \quad y'|_{x=0} = 0.$
$y_1 = 1,\ y_2 = \operatorname{arc\,tan} x.$

662. $(1 - x)y'' + xy' - y = -(x - 1)^2e^x, \quad \lim\limits_{x\to-\infty} y = 0, \quad y|_{x=0} = 1;$
$y_1 = x,\ y_2 = e^x.$

663. $2x^2(2 - \ln x)y'' + x(4 - \ln x)y' - y = \dfrac{(2 - \ln x)^2}{\sqrt{x}},$
$\lim\limits_{x\to+\infty} y = 0; \quad y_1 = \ln x,\ y_2 = \sqrt{x}.$

664. $y'' + \dfrac{2}{x}y' - y = 4e^x, \quad \lim\limits_{x\to-\infty} y = 0,\ y'|_{x=-1} = -\dfrac{1}{e};$
$$y_1 = \frac{e^x}{x},\ y_2 = \frac{e^{-x}}{x}.$$

665. $x^3(\ln x - 1)y'' - x^2y' + xy = 2\ln x, \quad \lim\limits_{x\to+\infty} y = 0;$

$y_1 = x, y_2 = \ln x.$

666. $(x^2 - 2x)y'' + (2 - x^2)y' - 2(1 - x)y = 2(x - 1),$

$\lim\limits_{x\to+\infty} y = 1; \quad y_1 = x^2, y_2 = e^x.$

§13. METHOD OF ISOCLINES FOR DIFFERENTIAL EQUATIONS OF THE SECOND ORDER

The method of isoclines (see §2) can also be applied to the solution of certain equations of the second order. These are equations which can be reduced to equations of the first order, for example, equations of the form

$$\frac{d^2x}{dt^2} + f\left(\frac{dx}{dt}, x\right) = 0. \tag{1}$$

We introduce a new variable $v = \frac{dx}{dt}$. Then $\frac{d^2x}{dt^2} = v\frac{dv}{dx}$ and equation (1) takes the form

$$\frac{dv}{dx} = -\frac{f(v, x)}{v}. \tag{2}$$

This is an equation of the first order in which x is the independent variable and which can be solved by the method of isoclines.

We shall regard x as the displacement of some point of a system, and $\frac{dx}{dt} = v$ as its speed.

The plane of the variables x, v is called the *phase plane*. Thus equation (2) determines the speed as a function of the displacement.

Constructing the field of isoclines for equation (2), we can draw an integral curve if we are given a starting point (x_0, v_0). Such a graphical representation of the speed v as a function of the displacement x is called a *phase picture*. A curve in the x, v plane which represents this relation is called a *phase trajectory*. Instantaneous values of x and v are coordinates of a point of the phase trajectory. This point is called a *representative point*. In the course of time, the representative point describes the phase trajectory. We note that positive speed means that the displacement increases with time. In fact, by virtue of the substitution $v = \frac{dx}{dt}$, if $v > 0$ we have $\frac{dx}{dt} > 0$, which means that x increases as t increases. Thus, in the upper half of the phase plane, where $v > 0$, the representative point must move from left to right. Therefore motion along a phase trajectory is performed clockwise.

Example. Construct a trajectory in the phase plane for the equation

$$\frac{d^2x}{dt^2} + x = 0. \tag{3}$$

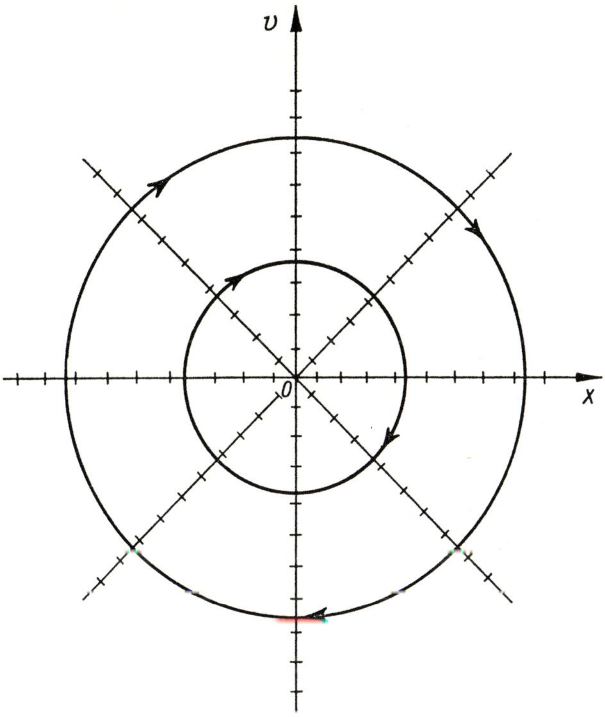

FIGURE 15

Solution. We put $\dfrac{dx}{dt} = v$. Equation (3) takes the form

$$v\frac{dv}{dx} + x = 0 \quad \text{or} \quad \frac{dv}{dx} = -\frac{x}{v}. \tag{4}$$

The equation of the isoclines for equation (4) is:

$$-\frac{x}{v} = k.$$

Drawing the isoclines corresponding to various values of k, we find that the phase trajectories are circles with centres at the point (0, 0) (Fig. 15).

We note that closed phase trajectories correspond to periodic motions. It is easy to see that in the case of equation (3) we do have periodic motion. Solving (3) by the methods explained above, we find that

$$x(t) = C_1 \cos t + C_2 \sin t.$$

Example. Construct the phase trajectories for the equation

$$\frac{d^2x}{dt^2} - \frac{dx}{dt} + x = 0. \tag{5}$$

Solution. We put $v = \frac{dx}{dt}$. Then equation (5) takes the form

$$\frac{dv}{dx} = \frac{v - x}{v}. \tag{6}$$

The equation of the isoclines is $\frac{v-x}{v} = k$. The phase trajectories have the form of unwinding spirals (Fig. 16). From the phase picture we can see that the motion is aperiodic, with amplitude increasing indefinitely with time.

Construct the phase trajectories for the following differential equations:

667. $\frac{d^2x}{dt^2} + \frac{dx}{dt} + x = 0.$

668. $\frac{d^2x}{dt^2} + 2\frac{dx}{dt} + 6x = 0.$

669. $\frac{d^2x}{dt^2} + 2\frac{dx}{dt} + x = 0.$

670. $\frac{d^2x}{dt^2} + \left(\frac{dx}{dt}\right)^2 + x = 0.$

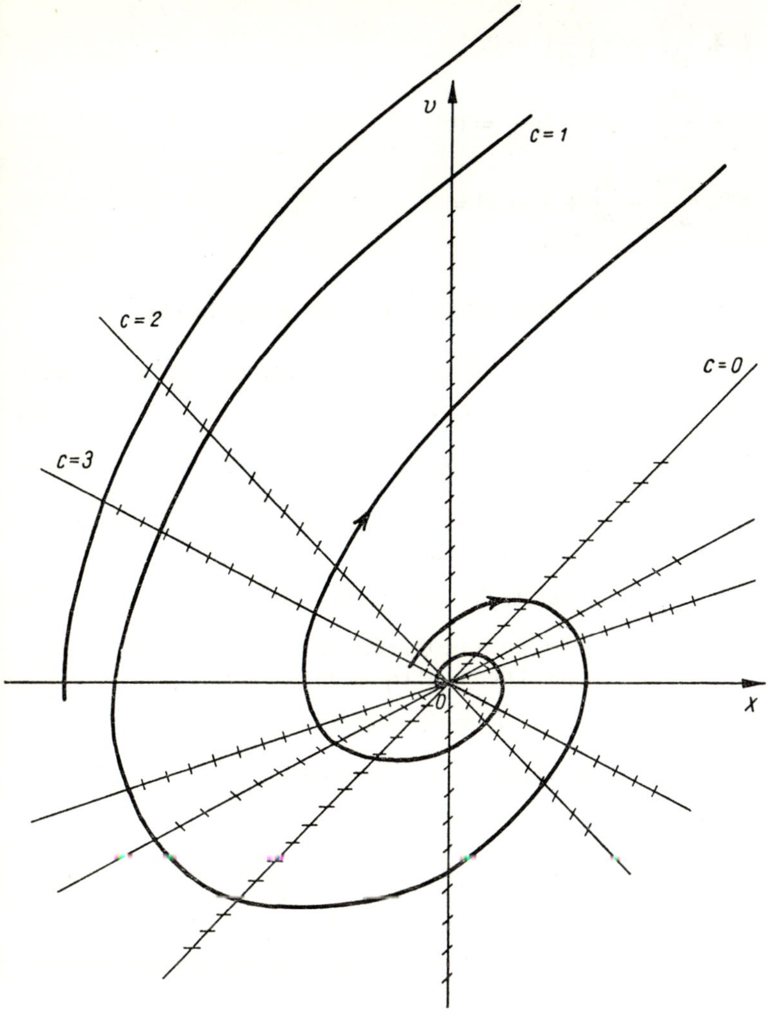

FIGURE 16

671. $\dfrac{d^2x}{dt^2} - 2\left(\dfrac{dx}{dt}\right)^2 + \dfrac{dx}{dt} - 2x = 0.$

672. $\dfrac{d^2x}{dt^2} - x \,.\, \exp\left(\dfrac{dx}{dt}\right) = 0 \ (\exp u \equiv e^u).$

673. $\dfrac{d^2x}{dt^2} + \exp\left(-\dfrac{dx}{dt}\right) - x = 0.$

674. $\dfrac{d^2x}{dt^2} + x\left(\dfrac{dx}{dt}\right)^2 = 0.$

675. $\dfrac{d^2x}{dt^2} + (x+2)\dfrac{dx}{dt} = 0.$

676. $\dfrac{d^2x}{dt^2} - \dfrac{dx}{dt} + x - x^2 = 0.$

§14. BOUNDARY VALUE PROBLEMS

We shall consider for simplicity an equation of the second order

$$y'' + p_1(x)y' + p_2(x)y = 0. \tag{1}$$

We shall assume that the coefficients $p_1(x)$ and $p_2(x)$ are continuous on a certain interval (a, b). Then every solution $y(x)$ of equation (1) is defined on the whole of this interval. In future, instead of equation (1), we shall consider the equation

$$[p(x)y']' - q(x)y = 0, \quad p(x) > 0. \tag{2}$$

Equations (1) and (2) can be transformed into each other.

A solution of the differential equation

$$[p(x)y']' - q(x)y = 0$$

is completely determined by the initial conditions

$$y(x_0) = y_0, \quad y'(x_0) = y_0'.$$

However, in many physical problems we have to look for solutions given in another way. For example, we might be given the problem: to find a solution of equation (2) which takes given values $y(a)$ and $y(b)$ at given points a and b. Usually in such cases the values of the solution are required only for values of x in (a, b). Thus the given values $y(a)$ and $y(b)$ occur at the ends of the interval, so a problem of this kind is called a boundary value problem. In future, we shall take as a basis the interval $(0, \pi)$ (the basic interval), which will not reduce the generality of the argument.

A very general form of boundary conditions for an equation of the second order is as follows:

$$\begin{aligned} h_0 y(0) + h_1 y'(0) &= A, \\ k_0 y(\pi) + k_1 y'(\pi) &= B, \end{aligned} \tag{3}$$

where h_0, h_1, k_0, k_1, A, B are given constants, and h_0, h_1, k_0, k_1 are not simultaneously zero.

If $A = B = 0$, the boundary conditions are called homogeneous. For example,

$$\left.\begin{array}{ll} (1) & y(0) = y(\pi) = 0. \\ (2) & y'(0) = h_0 y(0),\ y'(\pi) = -h_1 y(\pi);\quad h_0, h_1 > 0. \\ (3) & y'(0) = y'(\pi) = 0. \\ (4) & y(0) = y(\pi),\ y'(0) = y'(\pi). \end{array}\right\}. \qquad (3')$$

Generally speaking, boundary value problems are not always soluble, i.e. there does not always exist a solution which takes the required values at the ends of the interval. For example, the boundary value problem

$$y'' = 0, \quad y(0) - y(\pi) = 1, \quad y'(0) + y'(\pi) = 0$$

does not have a solution.

The problem

$$y'' + \lambda y = 0, \quad y(0) = y(\pi) = 0 \qquad (4)$$

has a non-zero solution only for integral values of $\sqrt{\lambda}$. In fact, from the general solution of the differential equation (4)

$$y = C_1 e^{\sqrt{-\lambda} x} + C_2 e^{-\sqrt{-\lambda} x}$$

it follows that the boundary conditions are satisfied if and only if $\lambda = n^2$, where n is an integer. The corresponding solutions are the functions $y_n = \sin nx$.

As is obvious from this example, if q in equation (2) is a function of a parameter λ, then under known conditions there exist values of the parameter for which the homogeneous boundary value problem given by equation (2) has a non-zero solution. These values of λ are called eigen-values, and the corresponding solutions of the boundary value problem are called eigen-functions. These are determined only to within an arbitrary constant multiplier. Thus for the boundary value problem $y'' + \lambda y = 0$, $y(0) = y(\pi) = 0$, the numbers 1^2, 2^2, $3^2, \ldots$ and the functions $\sin x$, $\sin 2x, \ldots$ are eigen-values and eigen-functions respectively.

As well as *simple* eigen-values, where to one eigen-value there corresponds one eigen-function (to within a constant multiplier), there exist *multiple* eigen-values, where to one eigen-value λ_0 there correspond two linearly independent eigen-functions.

To solve boundary value problems (for homogeneous linear differential equations) we proceed thus: we find the general solution of the given differential equation

$$y = C_1 y_1(x) + C_2 y_2(x) + \ldots + C_n y_n(x),$$

where

$$y_1(x), y_2(x), \ldots, y_n(x)$$

are linearly independent solutions. Then we require that this solution $y(x)$ should satisfy the given boundary conditions. This leads to a certain linear system of equations to determine $C_1, C_2, \ldots, C_n$. Solving this system, if possible, we find the solution of the given boundary value problem.

If the question of finding eigen-values arises, the condition for the existence of a non-zero solution of the system which determines $C_1, C_2, \ldots, C_n$ is the condition which determines the eigen-values. This is, in general, a certain transcendental equation for λ.

677. Under what conditions does the equation $y'' + \lambda y = 0$ have a non-zero solution satisfying the conditions:

$$(a)\ y'(0) = y'(\pi) = 0,$$

$$(b)\ \ y(0) = y(\pi), \quad y'(0) = y'(\pi)?$$

678. For what values of λ does the boundary value problem

$$y'' + \lambda y = 0,$$

$$y(0) = y(1) = 0$$

have the trivial solution $y \equiv 0$?

679. Which of the following boundary value problems is soluble:

$$(a)\ y'' - y = 0, \quad y(0) = 0, \quad y(2\pi) = 1,$$

$$(b)\ y'' + y = 0, \quad y(0) = 0, \quad y(2\pi) = 1?$$

680. Solve the boundary value problem

$$y'' + (\lambda - \omega^2)y = 0, \quad y(0) = y(1), \quad y'(0) = y'(1).$$

Consider the cases $\lambda - \omega^2 > 0$, $\lambda - \omega^2 = 0$, $\lambda - \omega^2 < 0$.

681. Find the solution of the equation $yy'' + (y')^2 + 1 = 0$ which passes through the points (0, 1) and (1, 2).

Solve the following boundary value problems:

682. $y''(x) = k^2 y(x)$, where $k^2 = \alpha^2 s^2$, $y(0) = v_1$, $y(x_0) = v_2$.

683. $y''(x) = \alpha^2 s y(x), \quad y(0) = v, \quad y'(x_0) = 0.$

684. $y''(x) - \alpha^2 s y(x) = 0,$

$$(a)\ y(0) = \frac{1}{s}, \quad y'(x_0) = 0,$$

$$(b)\ y'(0) = -\frac{1}{s}, \quad y(x_0) = 0.$$

685. $y''(x) = \alpha^2 s^2 y(x) + \alpha^2 g l, \quad y(0) = y(x_0) = 0.$

686. $y''(x) = \lambda^2 y(x), \quad (\lambda \neq 0),$

$$(a)\ y(0) = 0, \quad y(1) = \frac{1}{\lambda},$$

$$(b)\ y'(0) = 0, \quad y(1) = \frac{1}{\lambda},$$

$$(c)\ y(0) = 0, \quad y'(1) = \frac{1}{\lambda},$$

$$(d)\ y'(0) = 0, \quad y'(1) = \frac{1}{\lambda}.$$

687. $y^{\mathrm{IV}} - \lambda^4 y = 0, \quad y(0) = y''(0) = 0, \quad y(\pi) = y''(\pi) = 0.$

688. $y^{\mathrm{IV}} - \lambda^4 y = 0, \quad y(0) = y'(0) = 0, \quad y(\pi) = y''(\pi) = 0.$

689. $y^{\mathrm{IV}} - \lambda^4 y = 0, \quad y(0) = y'(0) = 0, \quad y(\pi) = y'(\pi) = 0.$

690. $y''' + \alpha y'' - \alpha^2 y' - \alpha^3 y = 0, \quad y(0) = -\dfrac{1}{\alpha},$

$$y'(0) = 1 + \frac{1}{\alpha}, \quad y(1) = 0.$$

691. $y^{\mathrm{IV}} - 2y''' + 2y'' - 2y' + y = \cos 2x, \quad y(0) = y(\pi) = \frac{1}{25},$
$y'(0) = \frac{2}{15}, \quad y'(\pi) = \frac{2}{25}.$

Note. The eigen-values in the problems considered above form an increasing numerical sequence. If the coefficients of the differential equation have a singular point on the boundary of the basic region, or if the basic region is infinite, for example, the whole number axis, then the spectrum, i.e. the set of eigen-values, can have a different structure. In particular, spectra can occur which contain all numbers in some interval of the value λ, the so-called continuous spectra.

For example, suppose that we are required to solve the equation $y'' + \lambda y = 0$ for the interval $-\infty < x < +\infty$ with the boundary condition: $y(x)$ is bounded at infinity. Obviously in this case any non-negative number λ is an eigen-value with eigen-functions $\sin \sqrt{\lambda}\, x$ and $\cos \sqrt{\lambda}\, x$.

In solving problems of mathematical physics which reduce to problems of determining eigen-values, we often obtain differential equations of the form

$$[p(x)y']' - q(x)y + \lambda \rho(x)y = 0,$$

but such that at the end points of the fundamental region the differential equation may have singularities, for example, the coefficient $p(x)$ may become zero. For these singular points, conditions arise from the very nature of the problem, for example, that the solution should be continuous or bounded or have a given order of infinity. These conditions play the part of boundary conditions. A typical example is the *Bessel equation*

$$(xy')' - \frac{n^2}{x}y + \lambda xy = 0, \tag{5}$$

which arises in various problems of mathematical physics. Here $p(x) \equiv x$ and the assumption made above that $p(x) > 0$ for the whole region $0 \leqslant x \leqslant 1$ is not satisfied here, since $p(0) = 0$. The point $x = 0$ is a singular point for the Bessel equation.

The requirement that the solution should be bounded at this point is a special form of boundary condition for the Bessel equation. So a boundary condition for the Bessel equation is as follows: find the solution of equation (5) which is bounded when $x = 0$ and, for example, becomes zero when $x = 1$.

Solve the boundary value problems:

692. $xy'' + y' = 0, \quad y(1) = \alpha y'(1),$

$y(x)$ bounded as $x \to 0$.

693. $x^2y^{IV} + 4xy''' + 2y'' = 0,$

$y(1) = y'(1) = 0, \quad y(x)$ bounded as $x \to 0$.

694. $x^3y^{IV} + 6x^2y''' + 6xy'' = 0,$

$y(1) = y'(1) = 0, \quad y(x)$ bounded as $x \to 0$.

§15. INTEGRATION OF DIFFERENTIAL EQUATIONS BY MEANS OF SERIES

This method is particularly suitable for linear differential equations. We shall illustrate it by applying it to an example of an equation of the second order.

Let a differential equation of the second order be given:

$$y'' + p(x)y' + q(x)y = 0. \tag{1}$$

We suppose that the coefficients $p(x)$ and $q(x)$ are given as series of increasing integral powers of x, so that equation (1) can be written in the form

$$y'' + (a_0 + a_1x + a_2x^2 + \ldots)y' + (b_0 + b_1x + b_2x^2 + \ldots)y = 0. \tag{2}$$

We look for a solution of this equation in the form of a power series

$$y = \sum_{k=0}^{\infty} c_k x^k. \tag{3}$$

Substituting this expression for y and its derivatives into (2), we obtain

$$\sum_{k=2}^{\infty} k(k-1)c_k x^{k-2} + \sum_{k=0}^{\infty} a_k x^k \sum_{k=1}^{\infty} k c_k x^{k-1} + \sum_{k=0}^{\infty} b_k x^k \sum_{k=0}^{\infty} c_k x^k = 0. \tag{4}$$

Multiplying the power series, collecting like terms and equating to zero the coefficients of the various powers of x on the left hand side of (4), we obtain a series of equations:

$$\begin{array}{l|l}
x^0 & 2.1\, c_2 + a_0c_1 + b_0c_0 = 0 \\
x^1 & 3.2\, c_3 + 2a_0c_2 + a_1c_1 + b_0c_1 + b_1c_0 = 0 \\
x^2 & 4.3\, c_4 + 3a_0c_3 + 2a_1c_2 + a_2c_1 + b_0c_2 + b_1c_1 + b_2c_0 = 0 \qquad (5) \\
\cdot & \cdot\quad\cdot\quad\cdot\quad\cdot\quad\cdot\quad\cdot\quad\cdot\quad\cdot\quad\cdot\quad\cdot\quad\cdot \\
\cdot & \cdot\quad\cdot\quad\cdot\quad\cdot\quad\cdot\quad\cdot\quad\cdot\quad\cdot\quad\cdot\quad\cdot\quad\cdot \\
\cdot & \cdot\quad\cdot\quad\cdot\quad\cdot\quad\cdot\quad\cdot\quad\cdot\quad\cdot\quad\cdot\quad\cdot\quad\cdot
\end{array}$$

Each of the equations (5) contains one more unknown coefficient than the previous one. The coefficients c_0 and c_1 remain arbitrary, and play the part of arbitrary constants.

The first of the equations (5) gives c_2, the second gives c_3, the third gives c_4, and so on, and in general if $c_0, c_1, \ldots, c_{k+1}$ are known then c_{k+2} can be determined from the $(k+1)$-th equation.

In practice, it is convenient to proceed as follows. According to the scheme described above, we determine two solutions $y_1(x)$ and $y_2(x)$, where for $y_1(x)$ we choose $c_0 = 1$ and $c_1 = 0$, and for $y_2(x)$ we choose $c_0 = 0$ and $c_1 = 1$, which are equivalent to the initial conditions:

$$\left.\begin{aligned} y_1(0) = 1, \quad y_1'(0) = 0 \\ y_2(0) = 0, \quad y_2'(0) = 1 \end{aligned}\right\}. \tag{6}$$

Any solution of equation (1) is a linear combination of the solutions $y_1(x)$ and $y_2(x)$.

If the initial conditions take the form $y(0) = A$, $y'(0) = B$, then obviously

$$y = Ay_1(x) + By_2(x).$$

Finally we give (without proof) a theorem about the existence of a solution of equation (1) in the form of a series (3).

Theorem. If the series

$$p(x) = \sum_{k=0}^{\infty} a_k x^k \quad \text{and} \quad q(x) = \sum_{k=0}^{\infty} b_k x^k$$

are convergent when $|x| < R$, then the power series (3) constructed by the method indicated above also converges for these values of x and is a solution of equation (1).

In particular, if $p(x)$ and $q(x)$ are polynomials in x, then the series (3) will converge for all values of x.

Example. Find the solution of the equation

$$y'' - xy' - 2y = 0 \tag{1}$$

in the form of a power series.

Solution. We look for $y_1(x)$ in the form

$$y_1(x) = \sum_{k=0}^{\infty} c_k x^k. \tag{2}$$

Then

$$y_1'(x) = \sum_{k=1}^{\infty} k c_k x^{k-1}, \tag{3}$$

$$y_1''(x) = \sum_{k=2}^{\infty} (k-1)k c_k x^{k-2}. \tag{4}$$

Substituting (2), (3), (4) into (1), we obtain

$$\sum_{k=2}^{\infty}(k-1)kc_kx^{k-2} - \sum_{k=1}^{\infty}kc_kx^k - 2\sum_{k=0}^{\infty}c_kx^k = 0. \tag{5}$$

Collecting like terms and equating to zero the coefficients of the various powers of x, we obtain relations from which $c_2, c_3, c_4, \ldots$ can be found.

We assume for definiteness that

$$y_1(0) = 1, \quad y_1'(0) = 0. \tag{6}$$

Then, from (2) and (6),

$$c_0 = 1, \tag{7}$$

and from (3) and (6),

$$c_1 = 0. \tag{8}$$

So we have

x^0	$2c_2 - 2c_0 = 0,$ hence from (7) $c_2 = 1,$
x^1	$3.2c_3 - 1c_1 - 2c_1 = 0,$ hence from (8) $c_3 = 0,$
x^2	$4.3c_4 - 2c_2 - 2c_2 = 0,$ hence $c_4 = \frac{1}{3},$
x^3	$5.4c_5 - 3c_3 - 2c_3 = 0,$ hence $c_5 = 0,$
x^4	$6.5c_6 - 4c_4 - 2c_4 = 0,$ hence $c_6 = \dfrac{c_4}{5} = \dfrac{1}{3.5},$
.	. . . .
.	. . . .
.	. . . .

and consequently

$$y_1(x) = 1 + x^2 + \tfrac{1}{3}x^4 + \tfrac{1}{15}x^6 + \ldots \tag{9}$$

Similarly, if we take

$$y_2(x) = \sum_{k=0}^{\infty} A_kx^k \tag{10}$$

and the initial conditions

$$y_2(0) = 0, \quad y_2'(0) = 1, \tag{11}$$

we obtain

$$A_0 = 0, \quad A_1 = 1, \tag{12}$$

and substituting (10) into (1) we find that

$$\sum_{k=0}^{\infty} k(k-1)A_k x^{k-2} - \sum_{k=1}^{\infty} (k+2)A_k x^k = 0$$

$$\begin{array}{c|ll}
x^0 & 2A_2 = 0, & A_2 = 0, \\
x^1 & 3.2A_3 - 3A_1 = 0, & A_3 = \frac{1}{2}, \\
x^2 & 4.3A_4 - 4A_2 = 0, & A_4 = 0, \\
x^3 & 5.4A_5 - 5A_3 = 0, & A_5 = \dfrac{1}{2.4}, \\
x^4 & 6.5A_6 - 6A_4 = 0, & A_6 = 0, \\
x^5 & 7.6A_7 - 7A_5 = 0, & A_7 = \dfrac{1}{2.4.6}, \\
\cdot & \cdot \quad \cdot \quad \cdot \quad \cdot & \cdot \quad \cdot \\
\cdot & \cdot \quad \cdot \quad \cdot \quad \cdot & \cdot \quad \cdot \\
\cdot & \cdot \quad \cdot \quad \cdot \quad \cdot & \cdot \quad \cdot
\end{array}$$

It is obvious that

$$A_{2k} = 0, \quad A_{2k+1} = \frac{1}{2.4.6\ldots(2k)} \quad (k = 1, 2, 3, \ldots).$$

So,

$$y_2(x) = x + \frac{x^3}{2} + \frac{x^5}{2.4} + \frac{x^7}{2.4.6} + \ldots = x\sum_{k=0}^{\infty} \frac{\left(\frac{x^2}{2}\right)^k}{k!} = xe^{\frac{x^2}{2}}. \quad (13)$$

The general solution of equation (1) has the form

$$y(x) = Ay_1(x) + By_2(x),$$

where $y_1(x)$ and $y_2(x)$ are given by formulae (9) and (13) respectively.

Expansion of a solution in a generalized power series.

Definition. A series of the form

$$x^\rho \sum_{k=0}^{\infty} c_k x^k \qquad (c_0 \neq 0), \qquad (14)$$

where ρ is a given number and the power series $\sum_{k=0}^{\infty} c_k x^k$ converges in some region $|x| < R$, is called a generalized power series.

If ρ is a non-negative integer, then a generalized power series (14) reduces to an ordinary power series.

The following theorem holds.

Theorem. If the point $x = 0$ is a singular point of equation (1), but the coefficients $p(x)$ and $q(x)$ of this equation are ratios of series to powers of x of degree not greater than 1 and 2 respectively:

$$p(x) = \frac{\sum_{k=0}^{\infty} a_k x^k}{x}, \quad q(x) = \frac{\sum_{k=0}^{\infty} b_k x^k}{x^2}, \tag{15}$$

where the series in the numerators converge in some region $|x| < R$, and the coefficients a_0, b_0 and b_1 are not simultaneously zero, then equation (1) has at least one solution in the form of a generalized power series

$$y = x^\rho \sum_{k=0}^{\infty} c_k x^k \qquad (c_0 \neq 0), \tag{16}$$

where the series which appears in this solution converges at least in the region $|x| < R$.

To determine the index ρ and the coefficients c_k, we must substitute the series (16) into equation (1), cancel x^ρ and equate to zero the coefficients of the various powers of x (the method of undetermined coefficients).

The number ρ is found from the so-called indicial equation

$$\rho(\rho - 1) + a_0\rho + b_0 = 0, \tag{17}$$

where

$$a_0 = \lim_{x\to 0} xp(x), \quad b_0 = \lim_{x\to 0} x^2 q(x). \tag{18}$$

Suppose that ρ_1 and ρ_2 are the roots of the indicial equation (17). We shall distinguish three cases.

1. If the difference $\rho_1 - \rho_2$ is not equal to an integer (or zero), then we can form two solutions of the form (16):

$$y_1 = x^{\rho_1} \sum_{k=0}^{\infty} c_k x^k \qquad (c_0 \neq 0),$$

$$y_2 = x^{\rho_2} \sum_{k=0}^{\infty} A_k x^k \qquad (A_0 \neq 0).$$

2. If the difference $\rho_1 - \rho_2$ is a positive integer, then we can, generally speaking, form only one series which is a solution of equation (1):

$$y_1 = x^{\rho_1} \sum_{k=0}^{\infty} c_k x^k. \tag{19}$$

3. If equation (17) has a double root $\rho_1 = \rho_2$, then it is also possible to form only one series, the solution (19).

It is clear that in the first case the solutions $y_1(x)$ and $y_2(x)$ are linearly independent (i.e. their ratio is not constant). In the second and third cases we have formed only one solution (19). We note without proof that if the difference $\rho_1 - \rho_2$ is a positive integer or zero then as well as the solution (19) there is a solution of the form

$$y_2 = Ay_1(x) \ln x + x^{\rho_2} \sum_{k=0}^{\infty} A_k x^k. \tag{20}$$

We see that now $y_2(x)$ contains an additional term of the form

$$Ay_1(x) \ln x,$$

where $y_1(x)$ is given in the form (19).

Note. The constant A in (20) can be equal to zero, and then for $y_2(x)$ we obtain an expression in the form of a generalized power series.

Example. Solve the equation

$$2x^2y'' + (3x - 2x^2)y' - (x + 1)y = 0. \tag{1}$$

Solution. We write (1) in the form

$$y'' + \frac{3x - 2x^2}{2x^2} y' - \frac{x + 1}{2x^2} y = 0 \tag{2}$$

or

$$y'' + \frac{3 - 2x}{2x} y' - \frac{x + 1}{2x^2} y = 0. \tag{3}$$

We look for a solution $y(x)$ of the form

$$y(x) = x^{\rho} \sum_{k=0}^{\infty} C_k x^k \qquad (C_0 \neq 0). \tag{4}$$

To find ρ we write down the indicial equation

$$\rho(\rho - 1) + a_0\rho + b_0 = 0,$$

where

$$a_0 = \lim_{x \to 0} \frac{3 - 2x}{2} = \frac{3}{2}, \quad b_0 = \lim_{x \to 0} \left(-\frac{x + 1}{2}\right) = -\frac{1}{2},$$

so

$$\rho(\rho - 1) + \tfrac{3}{2}\rho - \tfrac{1}{2} = 0$$

or

$$\rho^2 + \tfrac{1}{2}\rho - \tfrac{1}{2} = 0.$$

Hence

$$\rho_1 = \tfrac{1}{2}, \quad \rho_2 = -1. \tag{5}$$

According to the given rule, we take

$$y_1(x) = x^{1/2} \sum_{k=0}^{\infty} C_k x^k, \tag{6}$$

$$y_2(x) = \frac{1}{x} \sum_{k=0}^{\infty} A_k x^k. \tag{7}$$

In order to find the coefficients $C_0, C_1, \ldots, C_n, \ldots$ we substitute $y_1(x)$ and its derivatives $y_1'(x)$ and $y_1''(x)$ into equation (1). We have

$$y_1(x) = \sum_{k=0}^{\infty} C_k x^{k+\frac{1}{2}}, \quad y_1'(x) = \sum_{k=0}^{\infty} (k + \tfrac{1}{2}) C_k x^{k-\frac{1}{2}},$$

$$y_1''(x) = \sum_{k=0}^{\infty} (k + \tfrac{1}{2})(k - \tfrac{1}{2}) C_k x^{k-\frac{3}{2}},$$

$$2x^2 \sum_{k=0}^{\infty} (k^2 - \tfrac{1}{4}) C_k x^{k-\frac{3}{2}} + (3x - 2x^2) \sum_{k=0}^{\infty} (k + \tfrac{1}{2}) C_k x^{k-\frac{1}{2}}$$

$$- (x + 1) \sum_{k=0}^{\infty} C_k x^{k+\frac{1}{2}} = 0. \tag{8}$$

After some rearrangement (8) can be written as

$$\sum_{k=0}^{\infty} k(2k + 3) C_k x^{k+\frac{1}{2}} - \sum_{k=0}^{\infty} 2(k + 1) C_k x^{k+\frac{3}{2}} = 0. \tag{9}$$

Since we are looking for a solution for $x > 0$, we may cancel $x^{1/2}$ and obtain

$$\sum_{k=0}^{\infty} k(2k + 3) C_k x^k - \sum_{k=0}^{\infty} 2(k + 1) C_k x^{k+1} = 0. \tag{10}$$

Hence we obtain relations for determining the coefficients:

$$\left.\begin{array}{c|l}
x^1 & 1.5C_1 - 2.1C_0 = 0 \\
x^2 & 2.7C_2 - 2.2C_1 = 0 \\
x^3 & 3.9C_3 - 2.3C_2 = 0 \\
\cdot & \cdot \quad \cdot \quad \cdot \quad \cdot \quad \cdot \quad \cdot \quad \cdot \quad \cdot \\
\cdot & \cdot \quad \cdot \quad \cdot \quad \cdot \quad \cdot \quad \cdot \quad \cdot \quad \cdot \\
x^n & n(2n + 3)C_n - 2nC_{n-1} = 0 \\
\cdot & \cdot \quad \cdot \quad \cdot \quad \cdot \quad \cdot \quad \cdot \quad \cdot \quad \cdot
\end{array}\right\}. \tag{11}$$

Putting $C_0 = 1$ in the first of the relations (11), we obtain $C_1 = \frac{2}{5}$.

From the second, we have $C_2 = \dfrac{2^2}{5 \cdot 7}$.

From the third, $C_3 = \dfrac{2^3}{5 \cdot 7 \cdot 9}$ and so on.

It is easy to see that

$$C_n = \frac{2^n}{5 \cdot 7 \cdot 9 \ldots (2n+3)} \quad (n = 1, 2, 3, \ldots).$$

So

$$y_1(x) = x^{1/2}\left[1 + \sum_{k=0}^{\infty} \frac{(2x)^k}{5 \cdot 7 \cdot 9 \ldots (2k+3)}\right]. \tag{12}$$

Similarly, we find the coefficients A_k. It turns out that if $A_0 = 1$,

$$A_1 = 1, \quad A_2 = \frac{1}{2!}, \ldots, A_k = \frac{1}{k!},$$

so

$$y_2(x) = \frac{1}{x}\sum_{k=0}^{\infty} \frac{x^k}{k!} \quad \text{or} \quad y_2(x) = \frac{e^x}{x}. \tag{13}$$

The general solution of equation (1) is

$$y(x) = Ay_1(x) + By_2(x), \tag{14}$$

where A and B are arbitrary constants, and the functions $y_1(x)$ and $y_2(x)$ are given by formulae (12) and (13).

Example. Solve the Bessel equation

$$x^2y'' + xy' + (x^2 - p^2)y = 0, \tag{1}$$

where p is a given constant.

Solution. We write (1) in the form

$$y'' + \frac{1}{x}y' + \frac{x^2 - p^2}{x^2}y = 0. \tag{2}$$

Here

$$p(x) = \frac{1}{x}, \quad q(x) = \frac{x^2 - p^2}{x^2},$$

(see formulae (15) on p. 129), so

$$\left.\begin{aligned} a_0 &= \lim_{x \to 0} xp(x) = 1 \\ b_0 &= \lim_{x \to 0} x^2q(x) = -p^2 \end{aligned}\right\} \text{(see formulae (18)).}$$

The indicial equation for ρ is

$$\rho(\rho-1)+1.\rho-p^2=0 \quad \text{or} \quad \rho^2-p^2=0. \tag{3}$$

The roots of equation (3) are

$$\left.\begin{array}{l}\rho_1=p\\ \rho_2=-p\end{array}\right\}. \tag{4}$$

We look for a first particular solution of the Bessel equation (1) in the form of a generalized power series

$$y=x^p\sum_{k=0}^{\infty}C_kx^k. \tag{5}$$

Substituting y, y' and y'' into equation (1) we obtain:

$$x^2\sum_{k=0}^{\infty}C_k(k+p)(k+p-1)x^{k+p-2}+x\sum_{k=0}^{\infty}C_k(k+p)x^{k+p-1} \\ +(x^2-p^2)\sum_{k=0}^{\infty}C_kx^{k+p}=0$$

or, after a simple rearrangement and cancellation of x^p:

$$\sum_{k=0}^{\infty}[(k+p)^2-p^2]C_kx^k+\sum_{k=0}^{\infty}C_kx^{k+2}=0. \tag{6}$$

Hence, equating to zero the coefficients of the various powers of x, we have:

$$\left.\begin{array}{l|l}
x^0 & (p^2-p^2)C_0=0,\\
x^1 & [(1+p)^2-p^2]C_1=0,\\
x^2 & [(2+p)^2-p^2]C_2+C_0=0,\\
x^3 & [(3+p)^2-p^2]C_3+C_1=0,\\
x^4 & [(4+p)^2-p^2]C_4+C_2=0,\\
. & \dots\dots\dots\dots,\\
. & \dots\dots\dots\dots,\\
x^k & [(k+p)^2-p^2]C_k+C_{k-2}=0,\\
. & \dots\dots\dots\dots
\end{array}\right\} \tag{7}$$

The first of the relations (7) is satisfied for any value of the coefficient C_0.

From the second of the relations (7) we obtain $C_1=0$.

From the third:

$$C_2 = -\frac{C_0}{(2+p)^2 - p^2} = -\frac{C_0}{2^2(1+p)}.$$

From the fourth: $C_3 = 0$.
From the fifth:

$$C_4 = -\frac{C_2}{(4+p)^2 - p^2} = \frac{C_0}{2^4(1+p)(2+p) \, . \, 1 \, . \, 2}.$$

It is clear that all coefficients with odd suffixes are equal to zero:

$$C_{2k+1} = 0 \qquad (k = 0, 1, 2, \ldots). \tag{8}$$

The coefficients with even suffixes have the form:

$$C_{2k} = \frac{(-1)^k C_0}{2^{2k}(p+1)(p+2)\ldots(p+k)k!} \quad (k = 1, 2, \ldots). \tag{9}$$

To simplify the later calculations, we shall assume that

$$C_0 = \frac{1}{2^p \Gamma(p+1)}, \tag{10}$$

where $\Gamma(\nu)$ is Euler's gamma function, which is defined for all positive values of ν (and also for all complex numbers with positive real part) in the following way:

$$\Gamma(\nu) = \int_0^\infty e^{-x} x^{\nu-1} \, dx. \tag{11}$$

The gamma function has the following important properties:

1. $\Gamma(\nu + 1) = \nu\Gamma(\nu)$.
2. $\Gamma(1) = 1$.

If k is a positive integer,

3. $\Gamma(\nu + k + 1) = (\nu + 1)(\nu + 2) \ldots (\nu + k)\Gamma(\nu + 1)$.
4. $\Gamma(k + 1) = k!$

Using (10) and the properties of the gamma function, we may express the coefficient C_{2k} as:

$$C_{2k} = \frac{(-1)^k}{2^{2k}(p+1)(p+2)\ldots(p+k)\, k! \, . \, 2^p\Gamma(p+1)}$$

$$= \frac{(-1)^k}{2^{2k+p} \, . \, k!\Gamma(p+k+1)}$$

since, according to property 3, $(p+1)(p+2)\dots(p+k)\Gamma(p+1)$ is equal to $\Gamma(p+k+1)$. Now a particular solution of the Bessel equation, which we shall in future denote by J_p, takes the form

$$J_p(x) = \sum_{k=0}^{\infty} \frac{(-1)^k}{k!\Gamma(p+k+1)} \left(\frac{x}{2}\right)^{2k+p}. \tag{12}$$

In the literature, this function is usually called the *Bessel function of the first kind, of order p.*

We look for a second particular solution of the Bessel equation (1) of the form

$$y = x^{-p} \sum_{k=0}^{\infty} C_k x^k, \tag{13}$$

where p is the second root of the indicial equation (3). It is clear that this solution can be obtained from the solution (12) by replacing p by $-p$, since equation (1) contains p to an even power and does not change if p is replaced by $-p$. So

$$J_{-p}(x) = \sum_{k=0}^{\infty} \frac{(-1)^k}{k!\Gamma(k+1-p)} \left(\frac{x}{2}\right)^{2k-p}. \tag{14}$$

This function is called the Bessel function of the first kind, of order $-p$.

If p is *not an integer*, the solutions $J_p(x)$ and $J_{-p}(x)$ are linearly independent, since their expansions in series begin with different powers of x and consequently a linear combination $\alpha_1 J_p(x) + \alpha_2 J_{-p}(x)$ can be identically zero only if $\alpha_1 = \alpha_2 = 0$.

If p is *an integer*, then $J_p(x)$ and $J_{-p}(x)$ can be shown to be linearly dependent, and in fact it turns out that

$$J_{-n}(x) = (-1)^n J_n(x) \qquad (n \text{ integral}). \tag{15}$$

So if p is an integer, instead of $J_{-p}(x)$ we must look for another solution which is linearly independent of $J_p(x)$. To do this we introduce a new function

$$Y_p(x) = \frac{J_p(x)\cos p\pi - J_{-p}(x)}{\sin p\pi}, \tag{16}$$

assuming at first that p is not an integer.

It is obvious that the function $Y_p(x)$ so defined is a solution of equation (1) (since it is a linear combination of the particular solutions $J_p(x)$ and $J_{-p}(x)$).

Taking the limit of (16) as p tends to *an integer*, we obtain a particular solution $Y_p(x)$ linearly independent of $J_p(x)$ and defined for integral values of p.

The function $Y_p(x)$ introduced here is called a *Bessel function of the second kind, of order p*, or a *Weber* function. Thus for any value of p, fractional or integral, we can form a *fundamental system of solutions* of the Bessel equation (1). Hence it follows that the general solution of equation (1) can be expressed in the form

$$y = AJ_p(x) + BY_p(x), \tag{17}$$

where A and B are arbitrary constants.

In the case where p is not an integer, it is obvious that the general solution of the Bessel equation can be taken in the form

$$y = \alpha_1 J_p(x) + \alpha_2 J_{-p}(x), \tag{18}$$

where α_1 and α_2 are arbitrary constants.

Note 1. An equation which occurs frequently,

$$x^2y'' + xy' + (k^2x^2 - p^2)y = 0, \tag{19}$$

where k is some constant ($k \neq 0$), can be reduced to the Bessel equation

$$\xi^2 \frac{d^2y}{d\xi^2} + \xi \frac{dy}{d\xi} + (\xi^2 - p^2)y = 0 \tag{20}$$

by means of the substitution $\xi = kx$.

The general solution of equation (20) (if p is not integral) is

$$y = \alpha_1 J_p(\xi) + \alpha_2 J_{-p}(\xi),$$

and then the general solution of equation (19) takes the form:

$$y = \alpha_1 J_p(kx) + \alpha_2 J_{-p}(kx),$$

and if p is integral

$$y = \alpha_1 J_p(kx) + \alpha_2 Y_p(kx).$$

Note 2. An extensive class of equations of the form

$$t^2 \frac{d^2u}{dt^2} + at \frac{du}{dt} + (b + ct^m)u = 0, \tag{21}$$

where a, b, c and m are constants (c and m non-zero) can be reduced to the Bessel equation by introducing a new variable x and a new function y defined by the formulae

$$y = t^\alpha u, \quad x = \gamma t^\beta. \tag{22}$$

Integrate the following differential equations by means of series:

695. $y' - 2xy = 0, \quad y(0) = 1.$

696. $4xy'' + 2y' + y = 0.$

697. $(1 + x)y' - ny = 0.$

698. $9x(1 - x)y'' - 12y' + 4y = 0.$

699. $y'' + xy' + y = 0.$

700. $y'' - xy' + y - 1 = 0, \quad y(0) = y'(0) = 0.$

In problems **701–705**, find six terms of the expansion of $y(x)$:

701. $y'' - (1 + x^2)y = 0, \quad y(0) = -2, \quad y'(0) = 2.$

702. $y'' = x^2y - y', \quad y(0) = 1, \quad y'(0) = 0.$

703. $y'' - ye^x = 0.$

704. $y' = x^2 + y^2, \quad y(0) = 0.$

705. $y' = e^y + xy, \quad y(0) = 0.$

Find the general solutions of the following Bessel equations:

706. $x^2y'' + xy' + (4x^2 - \frac{1}{9})y = 0.$

707. $x^2y'' + xy' + (x^2 - \frac{1}{4})y = 0.$

708. $y'' + \dfrac{1}{x}y' + \dfrac{1}{9}y = 0.$

709. $y'' + \dfrac{1}{x}y' + 4y = 0.$

710. $x^2y'' - 2xy' + 4(x^4 - 1)y = 0.$

711. $xy'' + \frac{1}{2}y' + \frac{1}{4}y = 0.$

712. $y'' + \dfrac{5}{x}y' + y = 0.$

713. $y'' + \dfrac{3}{x}y' + 4y = 0.$

Demonstrate the truth of the following relations:

714. $J_p'(x) = J_{p-1}(x) - \dfrac{p}{x}J_p(x),$

715. $J_p'(x) = -J_{p+1}(x) + \dfrac{p}{x}J_p(x),$

716. $J_{p+1}(x) = \dfrac{2p}{x} J_p(x) - J_{p-1}(x).$

717. $J_2(x) = J_0''(x) - \dfrac{1}{x} J_0'(x).$

718. $J_2(x) - J_0(x) = 2J_0''(x).$

719. $J_3(x) + 3J_0'(x) + 4J_0'''(x) = 0.$

720. $x^2 J_p''(x) = (p^2 - p - x^2)J_p(x) + xJ_{p+1}(x).$

721. Verify that $\sqrt{x}J_1(2\sqrt{x})$ and $\sqrt{x}\,Y_1(2\sqrt{x})$ are solutions of the equation

$$xy'' + y = 0.$$

722. Show that $\sqrt[4]{x}J_{-\frac{1}{2}}(\sqrt{x})$ and $\sqrt[4]{x}J_{\frac{1}{2}}(\sqrt{x})$ are solutions of the equation

$$xy'' + \tfrac{1}{2}y' + \tfrac{1}{4}y = 0.$$

723. Show that

$$C_1 J_p(x) \int_0^x \frac{dt}{tJ_p^2(t)} + C_2 J_p(x)$$

is a solution of the Bessel equation.

§16. SYSTEMS OF LINEAR DIFFERENTIAL EQUATIONS WITH CONSTANT COEFFICIENTS

A *linear system* with constant coefficients is a system of the form

$$\frac{dx_i}{dt} = \sum_{j=1}^{n} a_{ij}x_j + f_i(t) \qquad (i = 1, 2, \ldots, n), \tag{1}$$

where a_{ij} are given numbers and $f_i(t)$ are given functions.

A linear system is called homogeneous if all the $f_i(t) \equiv 0$.

A solution of the system (1) on an interval (a, b) is a set of functions

$$x_1 = \phi_1(t), \quad x_2 = \phi_2(t), \ldots, x_n = \phi_n(t), \tag{2}$$

defined and continuously differentiable on the interval (a, b) if the functions (2) transform the equations of the system (1) into identities, valid for all values of t in (a, b).

The problem of finding a solution

$$x_1 = x_1(t), \quad x_2 = x_2(t), \ldots, x_n = x_n(t), \tag{3}$$

satisfying the initial conditions

$$x_1(t_0) = x_1^0, \quad x_2(t_0) = x_2^0, \ldots, x_n(t_0) = x_n^0 \tag{4}$$

is called the Cauchy problem.

We shall consider the four commonest methods of solving systems of linear differential equations with constant coefficients of the form (1).

1. Reduction of the system to one equation of the n-th order

The simplest way of integrating the system (1) is to reduce it to one equation of order n. The equation thus obtained will be linear with constant coefficients.

We illustrate this method by an example of a system of two equations:

$$\left.\begin{aligned} \frac{dx}{dt} &= ax + by + f(t) \quad (1,1) \\ \frac{dy}{dt} &= cx + dy + g(t) \quad (1,2) \end{aligned}\right\} \tag{1}$$

Here a, b, c, d are constant coefficients, $f(t)$, $g(t)$ are given functions, and $x(t)$ and $y(t)$ are unknown functions. From equation (1, 1) we find that

$$y = \frac{1}{b}\left(\frac{dx}{dt} - ax - f(t)\right). \tag{2}$$

Substituting this value for y into the second equation of the system, and the derivative of this value for dy/dt, we obtain an equation of the second order for $x(t)$

$$A\frac{d^2x}{dt^2} + B\frac{dx}{dt} + Cx + P(t) = 0,$$

where A, B, C are constants.

Hence we find that $x = x(t, C_1, C_2)$ and, substituting this value for x and also dx/dt, into (2), we find y.

Example. Integrate the system of equations

$$\left.\begin{aligned} \frac{dx}{dt} &= y + 1 \quad (3,1) \\ \frac{dy}{dt} &= x + 1 \quad (3,2) \end{aligned}\right\} \tag{3}$$

From (3, 1) we find that

$$y = \frac{dx}{dt} - 1. \tag{4}$$

Substituting (4) into (3, 2), we obtain a linear differential equation of the second order with constant coefficients

$$\frac{d^2x}{dt^2} - 1 - x = 0. \tag{5}$$

The general solution of equation (5) is

$$x = C_1e^t + C_2e^{-t} - 1. \tag{6}$$

Substituting the derivative of this with respect to t into (4), we obtain:

$$y = C_1e^t - C_2e^{-t} - 1.$$

The general solution of the system (3) is:

$$\left.\begin{aligned} x &= C_1e^t + C_2e^{-t} - 1 \\ y &= C_1e^t - C_2e^{-t} - 1 \end{aligned}\right\}$$

2. Euler's method of integrating systems of homogeneous linear differential equations with constant coefficients

We shall consider this method by applying it to a system of three linear differential equations:

$$\left.\begin{aligned} \frac{dx}{dt} &= ax + by + cz \\ \frac{dy}{dt} &= a_1x + b_1y + c_1z \\ \frac{dz}{dt} &= a_2x + b_2y + c_2z \end{aligned}\right\} \tag{1}$$

We look for a solution of the system (1) in the form

$$x = \lambda e^{rt}, \quad y = \mu e^{rt}, \quad z = \nu e^{rt}, \tag{2}$$

where λ, μ, ν and r are constants.

Substituting (2) into (1) and cancelling e^{rt}, we obtain a system of equations to determine λ, μ and ν:

$$\left.\begin{aligned} (a - r)\lambda + b\mu + c\nu &= 0 \\ a_1\lambda + (b_1 - r)\mu + c_1\nu &= 0 \\ a_2\lambda + b_2\mu + (c_2 - r)\nu &= 0 \end{aligned}\right\}\cdot \tag{3}$$

From this system it is possible to obtain non-zero values of λ, μ and ν only if the determinant

$$\Delta = \begin{vmatrix} a - r & b & c \\ a_1 & b_1 - r & c_1 \\ a_2 & b_2 & c_2 - r \end{vmatrix} = 0. \tag{4}$$

Equation (4), the so-called characteristic equation, is cubic in r.

(*a*) Suppose that r_1, r_2 and r_3, the roots of the characteristic equation, are real and distinct. Substituting r_1 for r in equation (3), and solving the system (3), we obtain numbers λ_1, μ_1 and ν_1. Then we substitute into (3) $r = r_2$ and obtain numbers λ_2, μ_2, ν_2 and finally for $r = r_3$ we obtain λ_3, μ_3, ν_3. Corresponding to the three sets of numbers λ, μ and ν we obtain three systems of particular solutions:

$$\left.\begin{aligned} x_1 &= \lambda_1 e^{r_1 t}, & y_1 &= \mu_1 e^{r_1 t}, & z_1 &= \nu_1 e^{r_1 t} \\ x_2 &= \lambda_2 e^{r_2 t}, & y_2 &= \mu_2 e^{r_2 t}, & z_2 &= \nu_2 e^{r_2 t} \\ x_3 &= \lambda_3 e^{r_3 t}, & y_3 &= \mu_3 e^{r_3 t}, & z_3 &= \nu_3 e^{r_3 t} \end{aligned}\right\} . \tag{5}$$

The general solution of the system (1) has the form

$$\left.\begin{aligned} x &= C_1 x_1 + C_2 x_2 + C_3 x_3 \\ y &= C_1 y_1 + C_2 y_2 + C_3 y_3 \\ z &= C_1 z_1 + C_2 z_2 + C_3 z_3 \end{aligned}\right\} . \tag{6}$$

Example. Solve the system:

$$\left.\begin{aligned} \frac{dx}{dt} &= 3x - y + z \\ \frac{dy}{dt} &= -x + 5y - z \\ \frac{dz}{dt} &= x - y + 3z \end{aligned}\right\} . \tag{7}$$

Solution. We form the characteristic equation

$$\begin{vmatrix} 3 - r & -1 & 1 \\ -1 & 5 - r & -1 \\ 1 & -1 & 3 - r \end{vmatrix} = 0 \tag{8}$$

or $r^3 - 11r^2 + 36r - 36 = 0$. Corresponding to the roots $r_1 = 2$, $r_2 = 3$, $r_3 = 6$, we have

$$\begin{aligned} &\lambda_1 = 1, \quad \mu_1 = 0, \quad \nu_1 = -1; \\ &\lambda_2 = 1, \quad \mu_2 = 1, \quad \nu_2 = 1; \\ &\lambda_3 = 1, \quad \mu_3 = -2, \quad \nu_3 = 1. \end{aligned}$$

We write out the particular solutions

$$\left.\begin{aligned} &x_1 = e^{2t}, \quad y_1 = 0, \quad z = -e^{2t} \\ &x_1 = e^{3t}, \quad y_2 = e^{3t}, \quad z_2 = e^{3t} \\ &x_3 = e^{6t}, \quad y_3 = -2e^{6t}, \quad z_3 = e^{6t} \end{aligned}\right\} \tag{9}$$

The answer is

$$\left.\begin{aligned} x &= C_1e^{2t} + C_2e^{3t} + C_3e^{6t} \\ y &= C_2e^{3t} - 2C_3e^{6t} \\ z &= -C_1e^{2t} + C_2e^{3t} + C_3e^{6t} \end{aligned}\right\}. \tag{10}$$

(*b*) We now consider the case where the roots of the characteristic equation are complex.

Example. Solve the system

$$\left.\begin{aligned} \frac{dx}{dt} &= x - 5y \\ \frac{dy}{dt} &= 2x - y \end{aligned}\right\}. \tag{11}$$

Solution. We write down the system for determining λ and μ:

$$\left.\begin{aligned} (1 - r)\lambda - 5\mu &= 0 \\ 2\lambda - (1 + r)\mu &= 0 \end{aligned}\right\}. \tag{12}$$

The characteristic equation

$$\begin{vmatrix} 1 - r & -5 \\ 2 & -1 - r \end{vmatrix} = 0$$

has roots $r_1 = 3i$, $r_2 = -3i$. Substituting $r_1 = 3i$ into (12), we obtain two equations to determine λ_1 and μ_1:

$$\left.\begin{aligned} (1 - 3i)\lambda_1 - 5\mu_1 &= 0 \\ 2\lambda_1 - (1 + 3i)\mu_1 &= 0 \end{aligned}\right\},$$

of which one is a consequence of the other (by virtue of the fact that the determinant of the system (12) is zero).

We can choose $\lambda_1 = 5$, $\mu_1 = 1 - 3i$, then the first particular solution can be written as

$$x_1 = 5e^{3it}, \quad y_1 = (1 - 3i)e^{3it}. \tag{13}$$

Similarly, substituting into (12) the root $r_2 = -3i$, we find the second particular solution:

$$x_2 = 5e^{-3it}, \quad y_2 = (1 + 3i)e^{-3it}. \tag{14}$$

We go over to a new fundamental system of solutions

$$\left.\begin{aligned} \tilde{x}_1 &= \frac{x_1 + x_2}{2}, & \tilde{x}_2 &= \frac{x_1 - x_2}{2i} \\ \tilde{y}_1 &= \frac{y_1 + y_2}{2}, & \tilde{y}_2 &= \frac{y_1 - y_2}{2i} \end{aligned}\right\} \tag{15}$$

Using the well-known formula of Euler

$$e^{\pm\alpha it} = \cos \alpha t \pm i \sin \alpha t, \tag{16}$$

where α is a real number, we obtain from (14) and (15)

$$\tilde{x}_1 = 5 \cos 3t, \qquad \tilde{x}_2 = 5 \sin 3t;$$

$$\tilde{y}_1 = \cos 3t + 3 \sin 3t, \quad \tilde{y}_2 = \sin 3t - 3 \cos 3t.$$

The general solution of the system (11) is

$$x = C_1\tilde{x}_1 + C_2\tilde{x}_2 = 5C_1 \cos 3t + 5C_2 \sin 3t,$$

$$y = C_1\tilde{y}_1 + C_2\tilde{y}_2 = C_1(\cos 3t + 3 \sin 3t) + C_2(\sin 3t - 3 \cos 3t).$$

Note. Having found the first particular solution (13), we could have written down at once the general solution of the system (11), by using the formulae

$$\left.\begin{aligned} x &= C_1 \operatorname{Re} x_1 + C_2 \operatorname{Im} x_1 \\ y &= C_1 \operatorname{Re} y_1 + C_2 \operatorname{Im} y_1 \end{aligned}\right\}, \tag{17}$$

where $\operatorname{Re} z$ and $\operatorname{Im} z$ are the real and imaginary parts of the complex number z, i.e. if $z = a + bi$, then $\operatorname{Re} z = a$, $\operatorname{Im} z = b$.

(*c*) The case of a multiple root.

Example. Solve the system

$$\left.\begin{aligned} \frac{dx}{dt} &= 2x + y & (18, 1) \\ \frac{dy}{dt} &= 4y - x & (18, 2) \end{aligned}\right\} \tag{18}$$

Solution. The characteristic equation

$$\begin{vmatrix} 2-r & 1 \\ -1 & 4-r \end{vmatrix} = 0$$

or $r^2 - 6r + 9 = 0$ has a double root $r_1 = r_2 = 3$. We look for a solution of the form

$$\left.\begin{aligned} x &= (\lambda_1 + \mu_1 t)e^{3t} \\ y &= (\lambda_2 + \mu_2 t)e^{3t} \end{aligned}\right\}. \tag{19}$$

Substituting (19) into (18, 1), we obtain:

$$3(\lambda_1 + \mu_1 t) + \mu_1 = 2(\lambda_1 + \mu_1 t) + (\lambda_2 + \mu_2 t). \tag{20}$$

Equating the coefficients of the same powers of t on the left and right hand sides of (20), we obtain:

$$\left.\begin{aligned} 3\lambda_1 + \mu_1 &= 2\lambda_1 + \lambda_2 \\ 3\mu_1 &= 2\mu_1 + \mu_2 \end{aligned}\right\}, \tag{21}$$

and hence

$$\left.\begin{aligned} \lambda_2 &= \lambda_1 + \mu_1 \\ \mu_2 &= \mu_1 \end{aligned}\right\}. \tag{22}$$

The quantities λ_1 and μ_1 remain arbitrary. Denoting them by C_1 and C_2 respectively, we obtain the general solution of the system (18) in the form:

$$\left.\begin{aligned} x &= (C_1 + C_2 t)e^{3t} \\ y &= (C_1 + C_2 + C_2 t)e^{3t} \end{aligned}\right\}.$$

Note. It is easy to verify that if (19) is substituted into (18.2) we obtain the same result (22). In fact, from the equality

$$\mu_2 + 3(\lambda_2 + \mu_2 t) = 4(\lambda_2 + \mu_2 t) - (\lambda_1 + \mu_1 t)$$

we obtain two relations to determine λ_2 and μ_2 in terms of λ_1 and μ_1:

$$\left.\begin{aligned} \mu_2 + 3\lambda_2 &= 4\lambda_2 - \lambda_1 \\ 3\mu_2 &= 4\mu_2 - \mu_1 \end{aligned}\right\},$$

and hence

$$\lambda_2 = \lambda_1 + \mu_1, \quad \mu_2 = \mu_1.$$

3. Solution of systems of differential equations by finding integrable combinations

This method of integrating systems of differential equations (not necessarily linear) is as follows: by means of suitable arithmetic operations (for example, addition, subtraction, etc.) we form, from the equations of the given system, so-called integrable combinations, i.e. simply integrable equations of the form

$$F\left(t, u, \frac{du}{dt}\right) = 0, \tag{1}$$

where u is some function of the unknown functions

$$x_1(t), x_2(t), \ldots, x_n(t).$$

Example. Solve the system

$$\left.\begin{aligned} \frac{dx}{dt} &= \frac{y^2}{x} \\ \frac{dy}{dt} &= \frac{x^2}{y} \end{aligned}\right\} \tag{2}$$

Solution. We write the system (2) in the form

$$\left.\begin{aligned} x\,dx &= y^2\,dt \\ y\,dy &= x^2\,dt \end{aligned}\right\}. \tag{3}$$

Adding term by term, we obtain

$$x\,dx + y\,dy = (x^2 + y^2)\,dt$$

or

$$\frac{d(x^2 + y^2)}{x^2 + y^2} = 2\,dt,$$

and hence

$$\ln(x^2 + y^2) = 2t + \ln C_1.$$

Taking the exponential of each side, we have

$$x^2 + y^2 = C_1 e^{2t}. \tag{4}$$

Subtracting the second equation (3) term by term from the first, we obtain

$$x\,dx - y\,dy = (y^2 - x^2)\,dt$$

or

$$\frac{d(x^2 - y^2)}{x^2 - y^2} = -2\,dt,$$

and hence

$$x^2 - y^2 = C_2 e^{-2t}. \tag{5}$$

Solving (4) and (5) for x and y, we find the general solution of the system (2):

$$x = \sqrt{\tilde{C}_1 e^{2t} + \tilde{C}_2 e^{-2t}},$$

$$y = \sqrt{\tilde{C}_1 e^{2t} - \tilde{C}_2 e^{-2t}},$$

where for simplicity we put

$$\tilde{C}_1 = \tfrac{1}{2}C_1, \quad \tilde{C}_2 = \tfrac{1}{2}C_2.$$

Example. Find a particular solution of the system

$$\left.\begin{aligned} \frac{dx}{dt} &= 3x + 5y, && (6,1) \\ \frac{dy}{dt} &= -2x - 8y, && (6,2) \end{aligned}\right\} \tag{6}$$

which satisfies the initial conditions

$$x|_{t=0} = 2, \quad y|_{t=0} = 5. \tag{7}$$

Solution. Multiplying the first equation by 2 and adding it to the second, we obtain

$$\frac{d(2x+y)}{dt} = 2(2x+y),$$

and hence

$$2x + y = C_1 e^{2t},$$

$$y = C_1 e^{2t} - 2x. \tag{8}$$

Substituting (8) into (6, 1), we obtain a linear equation to determine x:

$$\frac{dx}{dt} = -7x + 5C_1 e^{2t}. \tag{9}$$

Hence

$$x = C_2 e^{-7t} + \tfrac{5}{9}C_1 e^{2t}. \tag{10}$$

Consequently,

$$y = -\tfrac{1}{9}C_1 e^{2t} - 2C_2 e^{-7t}. \tag{11}$$

The expressions (10) and (11) represent the general solution of the system (6).

In order to find the particular solution which satisfies the conditions (7), we must substitute for t, x and y in (10) and (11) the numbers 0, 2 and 5. We obtain a system of equations to determine C_1 and C_2:

$$\left.\begin{aligned} 2 &= C_2 + \tfrac{5}{9}C_1 \\ 5 &= -\tfrac{1}{9}C_1 - 2C_2 \end{aligned}\right\}, \tag{12}$$

and hence

$$C_1 = 9, \quad C_2 = -3.$$

The answer is:

$$\left.\begin{aligned} x &= 5e^{2t} - 3e^{-7t} \\ y &= -e^{2t} + 6e^{-7t} \end{aligned}\right\}.$$

4. The method of variation of parameters

We shall illustrate this method by an example of three non-homogeneous equations.

Suppose we are given a system

$$\left.\begin{aligned} x' + a_1x + b_1y + e_1z &= f_1(t) \\ y' + a_2x + b_2y + e_2z &= f_2(t) \\ z' + a_3x + b_3y + e_3z &= f_3(t) \end{aligned}\right\} \begin{matrix} (1, 1) \\ (1, 2) \\ (1, 3) \end{matrix} \tag{1}$$

We suppose that the general solution of the corresponding homogeneous system has already been found. Suppose it has the form:

$$\left.\begin{aligned} x &= C_1x_1 + C_2x_2 + C_3x_3 \\ y &= C_1y_1 + C_2y_2 + C_3y_3 \\ z &= C_1z_1 + C_2z_2 + C_3z_3 \end{aligned}\right\}. \tag{2}$$

We shall look for a solution of the non-homogeneous system (1) of the form

$$\left.\begin{aligned} x &= C_1(t)x_1 + C_2(t)x_2 + C_3(t)x_3 \\ y &= C_1(t)y_1 + C_2(t)y_2 + C_3(t)y_3 \\ z &= C_1(t)z_1 + C_2(t)z_2 + C_3(t)z_3 \end{aligned}\right\}. \tag{3}$$

where $C_1(t)$, $C_2(t)$, $C_3(t)$ are new unknown functions. We substitute (3) into (1); then equation (1, 1) takes the form

$$\begin{aligned} C_1'x_1 + C_2'x_2 + C_3'x_3 + C_1(x_1' + a_1x_1 + b_1y_1 + e_1z_1) \\ + C_2(x_2' + a_1x_2 + b_1y_2 + e_1z_2) \\ + C_3(x_3' + a_1x_3 + b_1y_3 + e_1z_3) = f_1(t). \end{aligned} \tag{4}$$

All the sums in the brackets are equal to zero (by virtue of the assumption (2)), so we have

$$C_1'x_1 + C_2'x_2 + C_3'x_3 = f_1(t). \tag{5}$$

Similarly, by substituting (3) into (1, 2) and (1, 3) we obtain

$$\left.\begin{aligned} C_1'y_1 + C_2'y_2 + C_3'y_3 &= f_2(t) \\ C_1'z_1 + C_2'z_2 + C_3'z_3 &= f_3(t) \end{aligned}\right\}. \tag{6}$$

The system of equations (5), (6), linear in C_1', C_2', C_3', has a solution, since its determinant

$$\Delta = \begin{vmatrix} x_1 & x_2 & x_3 \\ y_1 & y_2 & y_3 \\ z_1 & z_2 & z_3 \end{vmatrix} \neq 0$$

by virtue of the linear independence of the particular solutions of the corresponding homogeneous system.

Having found $C_1'(t)$, $C_2'(t)$, $C_3'(t)$, by integration we find $C_1(t)$, $C_2(t)$, $C_3(t)$ and hence the solution (3) of the non-homogeneous system (1).

Example. By the method of variation of parameters, solve the system

$$\left.\begin{aligned} \frac{dx}{dt} + 2x + 4y &= 1 + 4t \\ \frac{dy}{dt} + x - y &= \frac{3}{2}t^2 \end{aligned}\right\}. \tag{7}$$

Solution. First we solve the homogeneous system

$$\left.\begin{aligned} \frac{dx}{dt} + 2x + 4y &= 0 \\ \frac{dy}{dt} + x - y &= 0 \end{aligned}\right\}. \tag{8}$$

Differentiating the first equation with respect to t,

$$\frac{d^2x}{dt^2} + 2\frac{dx}{dt} + 4\frac{dy}{dt} = 0.$$

But from the second equation $\dfrac{dy}{dt} = y - x$, so we obtain

$$\frac{d^2x}{dt^2} + 2\frac{dx}{dt} + 4y - 4x = 0.$$

From the first equation $4y = -\dfrac{dx}{dt} - 2x$, so

$$\frac{d^2x}{dt^2} + 2\frac{dx}{dt} - \frac{dx}{dt} - 2x - 4x = 0$$

or

$$\frac{d^2x}{dt^2} + \frac{dx}{dt} - 6x = 0.$$

The general solution of this equation is

$$x = C_1e^{2t} + C_2e^{-3t}.$$

Since

$$y = -\frac{1}{2}x - \frac{1}{4}\frac{dx}{dt},$$

we obtain

$$y = -\,C_1e^{2t} + \tfrac{1}{4}C_2e^{-3t}.$$

Replacing C_2 by $4C_2$, we have

$$\left.\begin{aligned} x &= C_1e^{2t} + 4C_2e^{-3t} \\ y &= -C_1e^{2t} + C_2e^{-3t} \end{aligned}\right\}. \tag{9}$$

This is the general solution of the homogeneous system (8).

We now look for a solution of the non-homogeneous system (7) if the form

$$\left.\begin{aligned} x &= C_1(t)e^{2t} + 4C_2(t)e^{-3t} \\ y &= -C_1(t)e^{2t} + C_2(t)e^{-3t} \end{aligned}\right\}. \tag{10}$$

After substituting (10) into (7) we obtain

$$\left.\begin{aligned} C_1'(t)e^{2t} + 4C_2'(t)e^{-3t} &= 1 + 4t \\ -C_1'(t)e^{2t} + C_2'(t)e^{-3t} &= \tfrac{3}{2}t^2 \end{aligned}\right\}.$$

Hence

$$C_1'(t) = \frac{1 + 4t - 6t^2}{5}\,e^{-2t},$$

$$C_2'(t) = \frac{1 + 4t + \frac{3}{2}t^2}{5}\,e^{3t}.$$

Consequently

$$\left.\begin{aligned} C_1(t) &= \frac{t + 3t^2}{5} e^{-2t} + C_1 \\ C_2(t) &= \frac{t + \frac{1}{2}t^2}{5} e^{3t} + C_2 \end{aligned}\right\}, \tag{11}$$

where C_1 and C_2 are constants of integration.

Substituting these values for $C_1(t)$ and $C_2(t)$ into (10), we obtain the general solution of the system (7):

$$\left.\begin{aligned} x &= C_1e^{2t} + 4C_2e^{-3t} + t + t^2 \\ y &= -C_1e^{2t} + C_2e^{-3t} - \tfrac{1}{2}t^2 \end{aligned}\right\}. \tag{12}$$

We note that the functions $\phi_1(t) = t + t^2$ and $\phi_2(t) = -\frac{1}{2}t^2$ represent particular solutions of the non-homogeneous system (7).

Solve the following systems of linear differential equations with constant coefficients:

724. $\left.\begin{aligned} \frac{dx}{dt} &= 3 - 2y \\ \frac{dy}{dt} &= 2x - 2t \end{aligned}\right\}.$

725. $\left.\begin{aligned} \frac{dx}{dt} &= x - 2y \\ \frac{dy}{dt} &= x + 3y \end{aligned}\right\}.$

726. $\left.\begin{aligned} \frac{dx}{dt} + 3x + y &= 0 \\ \frac{dy}{dt} - x + y &= 0 \end{aligned}\right\}.$

727. $\left.\begin{aligned} \frac{dx}{dt} &= 3x - \frac{1}{2}y - 3t^2 - \frac{1}{2}t + \frac{3}{2} \\ \frac{dy}{dt} &= 2y - 2t - 1 \end{aligned}\right\}.$

728. $$\left.\begin{aligned} \frac{dx}{dt} &= -7x + y \\ \frac{dy}{dt} &= -5y - 2x \end{aligned}\right\}.$$

729. $$\left.\begin{aligned} \frac{dx}{dt} &= 2x - 9y \\ \frac{dy}{dt} &= x + 8y \end{aligned}\right\}.$$

730. $$\left.\begin{aligned} \frac{dx}{dt} &= y + z \\ \frac{dy}{dt} &= z + x \\ \frac{dz}{dt} &= x + y \end{aligned}\right\}.$$

731. $$\left.\begin{aligned} \frac{dx}{dt} &= y + z \\ \frac{dy}{dt} &= 3x + z \\ \frac{dz}{dt} &= 3x + y \end{aligned}\right\}.$$

732. $$\left.\begin{aligned} \frac{dx}{dt} &= 8y \\ \frac{dy}{dt} &= -2z \\ \frac{dz}{dt} &= 2x + 8y - 2z \end{aligned}\right\}.$$

733. $$\left.\begin{aligned} \frac{dx}{dt} &= 2x + y - 2z - t + 2 \\ \frac{dy}{dt} &= 1 - x \\ \frac{dz}{dt} &= x + y - z - t + 1 \end{aligned}\right\}.$$

734. $$\left.\begin{aligned} \frac{dx}{dt} &= -x + y + z + e^t \\ \frac{dy}{dt} &= x - y + z + e^{3t} \\ \frac{dz}{dt} &= x + y + z + 4 \end{aligned}\right\}.$$

735. $$\left.\begin{aligned} \frac{dx}{dt} &= x \cos t \\ 2\frac{dy}{dt} &= (e^t + e^{-t})y \end{aligned}\right\}.$$

736. $$\left.\begin{aligned} \frac{dx}{dt} &= e^t - y - 5x \\ \frac{dy}{dt} &= e^{2t} + x - 3y \end{aligned}\right\}, \quad x(0) = \frac{119}{900}, \quad y(0) = \frac{211}{900}.$$

737. $$\left.\begin{aligned} \frac{dx}{dt} &= 3x + 8y \\ \frac{dy}{dt} &= -3y - x \end{aligned}\right\}, \quad x(0) = 6, \quad y(0) = -2.$$

738. $$\left.\begin{aligned} \frac{dx}{dt} &= y \\ \frac{dy}{dt} &= -x \end{aligned}\right\}, \quad x(0) = y(0) = 1.$$

739. $$\left.\begin{aligned} \frac{dx}{dt} &= -4(x + y) \\ \frac{dx}{dt} + 4\frac{dy}{dt} &= -4y \end{aligned}\right\}, \quad x(0) = 1, \quad y(0) = 0.$$

740. $$\left.\begin{aligned} \frac{dx}{dt} &= 4x - 5y \\ \frac{dy}{dt} &= x \end{aligned}\right\}, \quad x(0) = 0, \quad y(0) = 1.$$

741. $$\left.\begin{aligned} \frac{dx}{dt} &= x + y + t \\ \frac{dy}{dt} &= x - 2y + 2t \end{aligned}\right\}, \quad x(0) = -\tfrac{7}{9}, \quad y(0) = -\tfrac{5}{9}.$$

742. $$\left.\begin{aligned} \frac{dx}{dt} &= x + 5y \\ \frac{dy}{dt} &= -3y - x \end{aligned}\right\}, \quad x(0) = -2, \quad y(0) = 1.$$

743. $$\left.\begin{aligned} \frac{dx}{dt} + 2\frac{dy}{dt} &= 17x + 8y \\ 13\frac{dx}{dt} &= 53x + 2y \end{aligned}\right\}, \quad x(0) = 2, \quad y(0) = -1.$$

744. $$\left.\begin{aligned} \frac{dx}{dt} &= y \\ \frac{dx}{dt} - \frac{dy}{dt} &= x + y \end{aligned}\right\}, \quad x(\pi) = -1, \quad y(\pi) = 0.$$

745. $$\left.\begin{aligned} \frac{dx}{dt} + \frac{dy}{dt} &= e^{-t} - y \\ 2\frac{dx}{dt} + \frac{dy}{dt} &= \sin t - 2y \end{aligned}\right\}, \quad x(0) = -2, \quad y(0) = 1.$$

746. $$\left.\begin{aligned} 2\frac{dx}{dt} &= 6x - y - 6t^2 - t + 3 \\ \frac{dy}{dt} &= 2y - 2t - 1 \end{aligned}\right\}, \quad x(0) = 2, y(0) = 3.$$

747. $$\left.\begin{aligned} \frac{dx}{dt} &= \frac{1}{y} \\ \frac{dy}{dt} &= \frac{1}{x} \end{aligned}\right\}.$$

748. $$\frac{dt}{my - nx} = \frac{dx}{nt - ly} = \frac{dy}{lx - mt}.$$

749. $\dfrac{dx}{y} = -\dfrac{dy}{x} = \dfrac{dp}{q} = -\dfrac{dq}{p}.$

750. $\dfrac{t\,dt}{y^2 - 2xy - x^2} = \dfrac{dx}{x + y} = \dfrac{dy}{x - y}.$

751. $\left.\begin{aligned} t\,dx &= (t - 2x)\,dt \\ t\,dy &= (tx + ty + 2x - t)\,dt \end{aligned}\right\}.$

752. $\dfrac{dt}{t^2 - x^2 - y^2} = \dfrac{dx}{2tx} = \dfrac{dy}{2ty}.$

753. $\dfrac{dt}{xt} = \dfrac{dx}{x^2} = \dfrac{dy}{txy - 2t^2}.$

754. $\dfrac{dt}{xy} = \dfrac{dx}{ty} = \dfrac{dy}{tx}.$

§17. THEORY OF STABILITY

1. Lyapunov stability

Suppose we have a system of differential equations

$$\frac{dx_i}{dt} = f_i(x_1, x_2, \ldots, x_n, t). \tag{1}$$

A solution $\phi_i(t)$ $(i = 1, 2, \ldots, n)$ of the system (1) which satisfies the initial conditions $\phi_i(t_0) = \phi_{i0}$ $(i = 1, 2, \ldots, n)$ is called *Lyapunov stable* if for any $\varepsilon > 0$ there exists a $\delta(\varepsilon) > 0$ such that for any solution $x_i(t)$ $(i = 1, 2, \ldots n)$ of the system (1) whose initial values satisfy the conditions

$$|x_i(t_0) - \phi_{i0}| < \delta \qquad (i = 1, 2, \ldots, n), \tag{2}$$

the inequality

$$|x_i(t) - \phi_i(t)| < \varepsilon \qquad (i = 1, 2, \ldots, n) \tag{3}$$

holds for *all* $t \geqslant t_0$.

If for arbitrarily small $\delta > 0$ even one solution $x_i(t)$ $(i = 1, 2, \ldots, n)$ does not satisfy the inequality (3), the solution $\phi_i(t)$ is called *unstable.*

If, in addition to the inequality (3) being satisfied under conditions (2), the conditions

$$\lim_{t\to\infty} |x_i(t) - \phi_i(t)| = 0, \qquad (i = 1, 2, \ldots, n), \tag{4}$$

are also satisfied, then the solution $\phi_i(t)$ $(i = 1, 2, \ldots, n)$ is called *asymptotically stable.*

The investigation of the stability of the solution $\phi_i(t)$ $(i = 1, 2, \ldots, n)$ of the system (1) can be reduced to the investigation of the stability of the null (trivial) solution $x_i \equiv 0$ $(i = 1, 2, \ldots, n)$ of a certain system analogous to the system (1)

$$\frac{dx_i}{dt} = F_i(x_1, x_2, \ldots, x_n, t) \qquad (i = 1, 2, \ldots, n), \tag{1'}$$

where $F_i(0, 0, \ldots, 0, t) \equiv 0$ $(i = 1, 2, \ldots, n)$.

The point $x_i \equiv 0$ $(i = 1, 2, \ldots, n)$ is called a *stationary point* of the system (1′).

In terms of a stationary point, the definition of stability and instability can be formulated thus: *a stationary point* $x_i \equiv 0$ $(i = 1, 2, \ldots, n)$ *is Lyapunov stable* if for any $\varepsilon > 0$ some $\delta > 0$ can be found such that any solution $x_i(t)$ $(i = 1, 2, \ldots, n)$ whose initial data $x_{i0} = x_i(t_0)$ $(i = 1, 2, \ldots, n)$ satisfy the condition

$$|x_{i0}| < \delta \qquad (i = 1, 2, \ldots, n) \tag{2'}$$

satisfies, for all $t \geqslant t_0$, the condition

$$|x_i(t)| < \varepsilon \qquad (i = 1, 2, \ldots, n). \tag{3'}$$

If, in addition to the inequality (3) being satisfied, the condition $\lim_{t\to\infty} |x_i(t)| = 0$ $(i = 1, 2, \ldots, n)$ is satisfied, *the stability is asymptotic.*

If for arbitrarily small $\delta > 0$ even one solution $x_i(t)$ $(i = 1, 2, \ldots, n)$ does not satisfy the condition (3′), then the *stationary point* of the system $x_i \equiv 0$ $(i = 1, 2, \ldots, n)$ is *unstable.*

Example. Working from the definition of Lyapunov stability, determine whether the solution of the system

$$\frac{dx}{dt} = -y, \quad \frac{dy}{dt} = x, \tag{1}$$

which satisfies the initial conditions $x(0) = 0$, $y(0) = 0$ is stable.

Solution. The solution of the system (1) which satisfies the given initial conditions is $x(t) \equiv 0$, $y(t) \equiv 0$. Any solution of this system which satisfies the conditions $x(0) = x_0$, $y(0) = y_0$ has the form

$$\begin{aligned} x(t) &= x_0 \cos t - y_0 \sin t, \\ y(t) &= x_0 \sin t + y_0 \cos t. \end{aligned} \tag{2}$$

We choose an arbitrary $\varepsilon > 0$ and show that there exists a $\delta(\varepsilon) > 0$ such that if

$$|x_0 - 0| < \delta$$
$$|y_0 - 0| < \delta$$

the inequalities

$$\begin{aligned} |x(t) - 0| &= |x_0 \cos t - y_0 \sin t| < \varepsilon, \\ |y(t) - 0| &= |x_0 \sin t + y_0 \cos t| < \varepsilon \end{aligned} \tag{3}$$

hold for all $t \geqslant 0$.

This also means, according to the definition, that the null solution $x(t) \equiv 0$, $y(t) \equiv 0$ of the system (1) is Lyapunov stable. We have, obviously,

$$\begin{aligned} |x_0 \cos t - y_0 \sin t| &\leqslant |x_0 \cos t| + |y_0 \sin t| \leqslant |x_0| + |y_0| \\ |x_0 \sin t + y_0 \cos t| &\leqslant |x_0 \sin t| + |y_0 \cos t| \leqslant |x_0| + |y_0| \end{aligned} \tag{4}$$

for all t. So, if $|x_0| + |y_0| < \varepsilon$, then *a fortiori*

$$\begin{aligned} |x_0 \cos t - y_0 \sin t| < \varepsilon \\ |x_0 \sin t + y_0 \cos t| < \varepsilon \end{aligned} \tag{5}$$

for all t.

Consequently if, for example, we choose $\delta(\varepsilon) = \frac{1}{2}\varepsilon$, then if $|x_0| < \delta$ and $|y_0| < \delta$ by virtue of (4) the inequality (5) holds for all $t \geqslant 0$, i.e. the null solution of the system (1) is in fact Lyapunov stable, but the stability is not asymptotic.

Starting from the definition of Lyapunov stability, investigate the stability of the solutions of the following equations and systems of equations:

755. $\dfrac{dx}{dt} = x + t, \quad x(0) = 1.$

756. $\dfrac{dx}{dt} = 2t(x + 1), \quad x(0) = 0.$

757. $\dfrac{dx}{dt} = -x + t^2, \quad x(1) = 1.$

758. $\dfrac{dx}{dt} = 2 + t, \quad x(0) = 1.$

759. $$\left.\begin{aligned}\frac{dx}{dt} &= x - 13y\\ \frac{dy}{dt} &= \frac{1}{4}x - 2y\end{aligned}\right\},\quad x(0) = y(0) = 0.$$

760. $$\left.\begin{aligned}\frac{dx}{dt} &= -x - 3y\\ \frac{dy}{dt} &= x - y\end{aligned}\right\},\quad x(0) = y(0) = 0.$$

2. Simplest types of stationary points

Suppose we have a system of two homogeneous linear differential equations with constant coefficients

$$\left.\begin{aligned}\frac{dx}{dt} &= a_{11}x + a_{12}y\\ \frac{dy}{dt} &= a_{21}x + a_{22}y\end{aligned}\right\}, \tag{1}$$

where

$$\Delta = \begin{vmatrix} a_{11} & a_{12} \\ a_{21} & a_{22} \end{vmatrix} \neq 0.$$

The point $x = 0$, $y = 0$, at which the right hand sides of the equations of the system (1) become zero, is called a *stationary point of the system* (1).

To investigate the stationary points of the system (1) we must form the characteristic equation

$$\begin{vmatrix} a_{11} - \lambda & a_{12} \\ a_{21} & a_{22} - \lambda \end{vmatrix} = 0 \tag{2}$$

and find its roots λ_1 and λ_2.

The following cases are possible:

I. The roots λ_1, λ_2 of the characteristic equation (2) are real and distinct:

(*a*) $\lambda_1 < 0$, $\lambda_2 < 0$. The stationary point is asymptotically stable (a stable node, Fig. 17),

(*b*) $\lambda_1 > 0$, $\lambda_2 > 0$. The stationary point is unstable (an unstable node, Fig. 18),

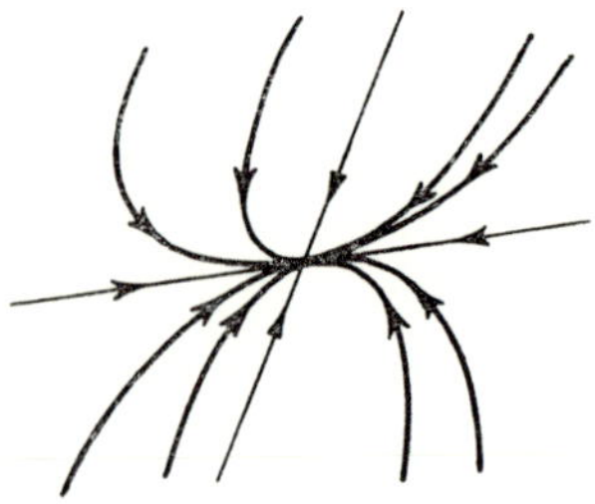

FIGURE 17

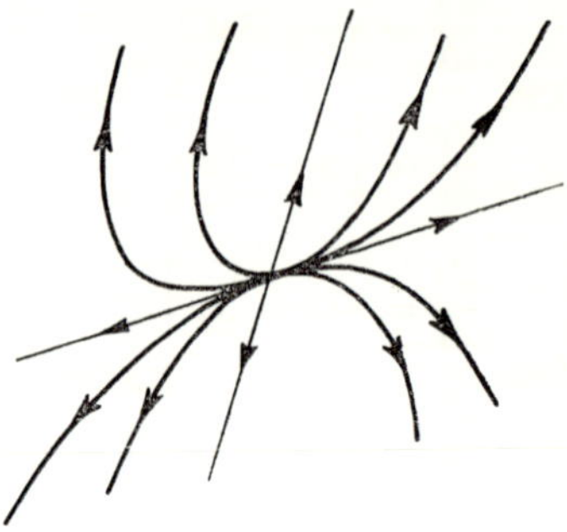

FIGURE 18

(*c*) $\lambda_1 > 0$, $\lambda_2 < 0$. The stationary point is unstable (a saddle point, Fig. 19).

II. The roots of the characteristic equation (2) are complex:

$$\lambda_1 = p + iq, \quad \lambda_2 = p - iq,$$

(*a*) $p < 0$, $q \neq 0$. The stationary point is asymptotically stable (a stable focus, Fig. 20),

(*b*) $p > 0$, $q \neq 0$. The stationary point is unstable (an unstable focus, Fig. 21),

(*c*) $p = 0$, $q \neq 0$. The stationary point is stable (a centre, Fig. 22).

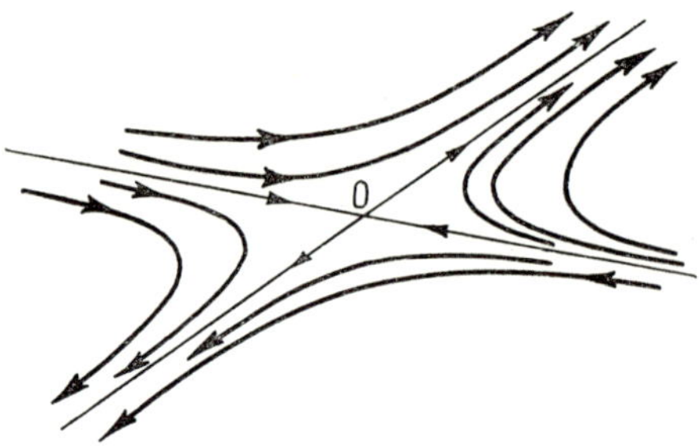

FIGURE 19

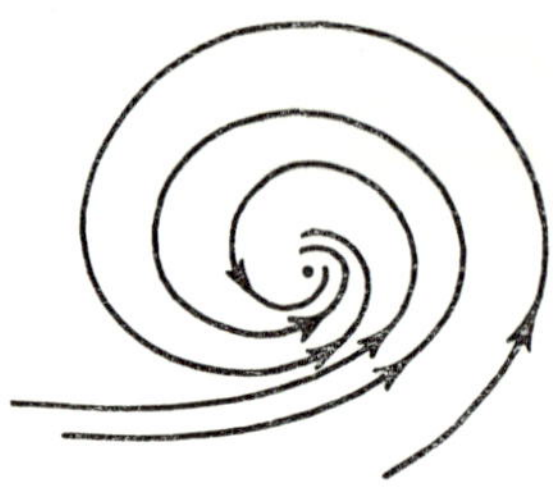

FIGURE 20

III. There is a double root $\lambda_1 = \lambda_2$:

(*a*) $\lambda_1 = \lambda_2 < 0$. The stationary point is asymptotically stable (a stable node, Figs. 23, 24).

(*b*) $\lambda_1 = \lambda_2 > 0$. The stationary point is unstable (an unstable node, Figs. 25, 26).

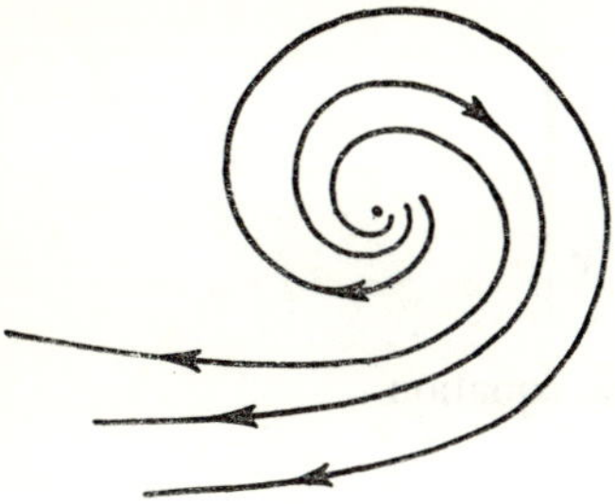

FIGURE 21

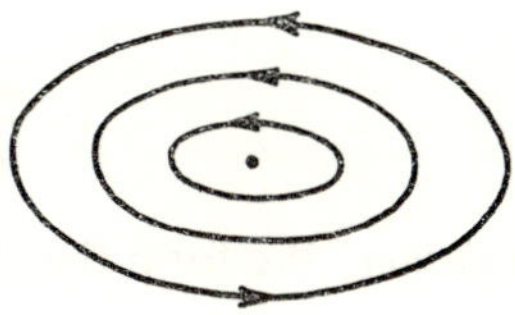

FIGURE 22

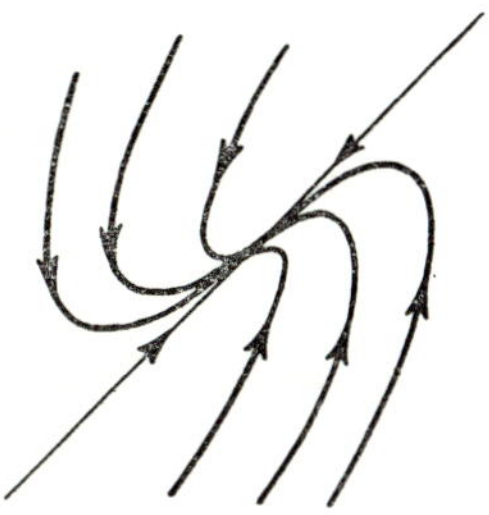

FIGURE 23

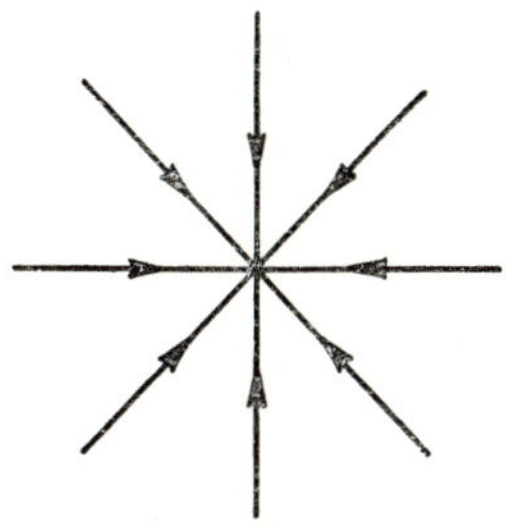

FIGURE 24

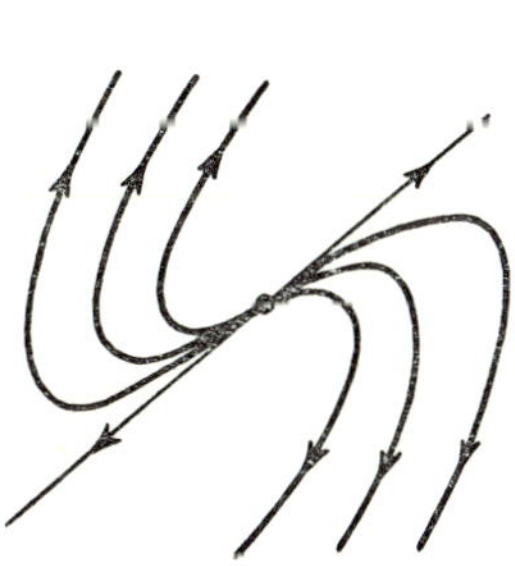

FIGURE 25

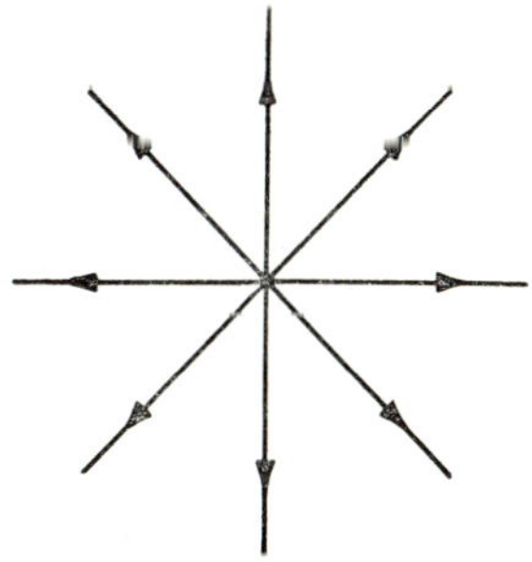

FIGURE 26

Example. Determine the character of the stationary point (0, 0) of the system

$$\left.\begin{aligned}\frac{dx}{dt} &= 5x - y\\ \frac{dy}{dt} &= 2x + y\end{aligned}\right\}.$$

Solution. We form the characteristic equation

$$\begin{vmatrix} 5-\lambda & -1 \\ 2 & 1-\lambda \end{vmatrix} = 0$$

or

$$\lambda^2 - 6\lambda + 7 = 0.$$

Its roots are $\lambda_1 = 3 + \sqrt{2} > 0$, $\lambda_2 = 3 - \sqrt{2} > 0$.

So the roots of the characteristic equation are real, distinct and positive. Consequently the stationary point (0, 0) is an unstable node.

Determine the character of the stationary points for the following systems of differential equations:

761. $$\left.\begin{aligned}\frac{dx}{dt} &= 3x + y\\ \frac{dy}{dt} &= -2x + y\end{aligned}\right\}.$$

762. $$\left.\begin{aligned}\frac{dx}{dt} &= -x + 2y\\ \frac{dy}{dt} &= x + y\end{aligned}\right\}.$$

763. $$\left.\begin{aligned}\frac{dx}{dt} &= -x + 3y\\ \frac{dy}{dt} &= -x + y\end{aligned}\right\}.$$

764. $$\left.\begin{aligned}\frac{dx}{dt} &= -2x - y\\ \frac{dy}{dt} &= 3x - y\end{aligned}\right\}.$$

765. $$\left.\begin{aligned}\frac{dx}{dt} &= 3x + 7y\\ \frac{dy}{dt} &= 2x + 5y\end{aligned}\right\}.$$

766. $$\left.\begin{aligned}\frac{dx}{dt} &= -2x + \frac{5}{7}y\\ \frac{dy}{dt} &= 7x - 3y\end{aligned}\right\}.$$

767. $$\left.\begin{aligned}\frac{dx}{dt} &= 3x - y\\ \frac{dy}{dt} &= x + y\end{aligned}\right\}.$$

768. $$\left.\begin{aligned}\frac{dx}{dt} &= -x - y\\ \frac{dy}{dt} &= x - 3y\end{aligned}\right\}.$$

769. $$\left.\begin{aligned}\frac{dx}{dt} &= 3x\\ \frac{dy}{dt} &= 3y\end{aligned}\right\}.$$

770. For what values of α is the stationary point (0, 0) of the system

$$\left.\begin{aligned}\frac{dx}{dt} &= -3x + \alpha y\\ \frac{dy}{dt} &= 2x + y\end{aligned}\right\}$$

stable?

Note. Suppose we have a system of homogeneous linear differential equations with constant coefficients

$$\frac{dx_i}{dt} = \sum_{j=1}^{n} a_{ij}x_j \qquad (i = 1, 2, \ldots, n) \qquad (n \geqslant 2). \tag{3}$$

For this there are analogous types of arrangement of the integral curves about the origin (which is called a generalized saddle point, generalized node, etc.). If all the roots of the characteristic equation of the system (3) have negative real part, then the stationary point $x_i \equiv 0$ $(i = 1, 2, \ldots, n)$ of the system (3) is asymptotically stable. If even one of the roots of the characteristic equation has positive real part, then the stationary point is unstable.

Example. Determine whether the stationary point (0, 0, 0) of the system

$$\left.\begin{aligned}\frac{dx}{dt} &= -x + z\\ \frac{dy}{dt} &= -2y - z\\ \frac{dz}{dt} &= y - z\end{aligned}\right\}.$$

is stable.

Solution. We form the characteristic equation

$$\begin{vmatrix} -1-\lambda & 0 & 1\\ 0 & -2-\lambda & -1\\ 0 & 1 & -1-\lambda \end{vmatrix} = 0$$

or $(1 + \lambda)(\lambda^2 + 3\lambda + 3) = 0$.

The roots of the characteristic equation

$$\lambda_1 = -1, \quad \lambda_{2,3} = \tfrac{1}{2}(-3 \pm i\sqrt{3})$$

have negative real part. Consequently the stationary point of the given system is asymptotically stable.

3. Stability to a first approximation

Suppose we have a system of differential equations

$$\frac{dx_i}{dt} = f_i(x_1, x_2, \ldots, x_n) \qquad (i = 1, 2, \ldots, n) \tag{1}$$

and let $x_i \equiv 0$ $(i = 1, 2, \ldots, n)$ be a stationary point of the system (1), i.e.

$$f_i(0, 0, \ldots, 0) = 0 \qquad (i = 1, 2, \ldots, n).$$

We shall assume that the functions $f_i(x_1, x_2, \ldots, x_n)$ are differentiable sufficiently often at the origin.

We expand the functions f_i in powers of the x_i in the neighbourhood of the origin, by Taylor's formula. We obtain

$$\frac{dx_i}{dt} = \sum_{j=1}^{n} a_{ij}x_j + R_i(x_1, x_2, \ldots, x_n) \qquad (i = 1, 2, \ldots, n), \tag{2}$$

where $a_{ij} = \dfrac{\partial f_i(0, 0, \ldots, 0)}{\partial x_j}$ and the R_i are terms of the second degree at least in $x_1, x_2, \ldots, x_n$.

Instead of the system (1), we consider the system

$$\frac{dx_i}{dt} = \sum_{j=1}^{n} a_{ij}x_j \qquad (i = 1, 2, \ldots, n) \qquad (a_{ij} = \text{const}), \tag{3}$$

which is called a first approximation to the system (1).

The following results are true:

1. If all the roots of the characteristic equation

$$\begin{vmatrix} a_{11} - \lambda & a_{12} & \ldots & a_{1n} \\ a_{21} & a_{22} - \lambda & \ldots & a_{2n} \\ \cdot & \cdot & \cdot & \cdot \\ a_{n1} & a_{n2} & \ldots & a_{nn} - \lambda \end{vmatrix} = 0 \tag{4}$$

have negative real parts, then the null solution $x_i \equiv 0$ $(i = 1, 2, \ldots, n)$ of the system (3) and the system (2) is asymptotically stable.

2. If even one root of the characteristic equation (4) has positive real part, then the null solution of the system (2) is unstable.

3. If the real parts of all the roots of the characteristic equation (4) are non-positive, where the real part of any one root is zero, then the investigation of stability to a first approximation is, generally speaking, impossible (the non-linear terms R_i begin to have an effect).

Example. Investigate the stability to a first approximation of the singular point $x = 0$, $y = 0$ of the system

$$\left.\begin{aligned} \dot{x} &= 2x + y - 5y^2 \\ \dot{y} &= 3x + y + \frac{x^3}{2} \end{aligned}\right\}, \qquad \left(\dot{x} = \frac{dx}{dt}, \quad \dot{y} = \frac{dy}{dt}\right). \tag{5}$$

Solution. The first approximation system is

$$\left.\begin{aligned} \dot{x} &= 2x + y \\ \dot{y} &= 3x + y \end{aligned}\right\}; \tag{6}$$

the non-linear terms satisfy the necessary condition that their degree is at least 2. We form the characteristic equation of the system (6):

$$\begin{vmatrix} 2 - \lambda & 1 \\ 3 & 1 - \lambda \end{vmatrix} = 0 \quad \text{or} \quad \lambda^2 - 3\lambda - 1 = 0. \tag{7}$$

The roots of the characteristic equation (7) $\lambda_1 = \frac{1}{2}(3 + \sqrt{13})$, $\lambda_2 = \frac{1}{2}(3 - \sqrt{13})$ are real and $\lambda_1 > 0$. Consequently the null solution $x = 0$, $y = 0$ of the system (5) is unstable.

Investigate the stability to a first approximation of the null solution $x = 0$, $y = 0$ of the following systems:

771. $$\left.\begin{aligned} \dot{x} &= x + 2y - \sin y^2 \\ \dot{y} &= -x - 3y + x(e^{\frac{x^2}{2}} - 1) \end{aligned}\right\}.$$

772. $$\left.\begin{aligned} \dot{x} &= -x + 3y + x^2 \sin y \\ \dot{y} &= -x - 4y + 1 - \cos y^2 \end{aligned}\right\}.$$

773. $$\left.\begin{aligned} \dot{x} &= -2x + 8 \sin^2 y \\ \dot{y} &= x - 3y + 4x^3 \end{aligned}\right\}.$$

774. $$\left.\begin{aligned} \dot{x} &= 3x - 22 \sin y + x^2 - y^3 \\ \dot{y} &= \sin x - 5y + e^{x^2} - 1 \end{aligned}\right\}.$$

775. $$\left.\begin{aligned}\dot{x} &= -10x + 4e^y - 4\cos y^2\\ \dot{y} &= 2e^x - 2 - y + x^4\end{aligned}\right\}.$$

776. $$\left.\begin{aligned}\dot{x} &= 7x + 2\sin y - y^4\\ \dot{y} &= e^x - 3y - 1 + \tfrac{5}{2}x^2\end{aligned}\right\}.$$

777. $$\left.\begin{aligned}\dot{x} &= -\tfrac{3}{2}x + \tfrac{1}{2}\sin 2y - x^3y\\ \dot{y} &= -y - 2x + x^4 - y^7\end{aligned}\right\}.$$

778. $$\left.\begin{aligned}\dot{x} &= \tfrac{5}{2}xe^x - 3y + \sin x^2\\ \dot{y} &= 2x + ye^{-\frac{y^2}{2}} - y^4\cos x\end{aligned}\right\}.$$

779. $$\left.\begin{aligned}\dot{x} &= \tfrac{3}{4}\sin x - 7y(1-y)^{1/3} + x^3\\ \dot{y} &= \tfrac{2}{3}x - 3y\cos y - 11y^5\end{aligned}\right\}.$$

780. $$\left.\begin{aligned}\dot{x} &= \tfrac{1}{4}(e^x - 1) - 9y + x^4\\ \dot{y} &= \tfrac{1}{5}x - \sin y + y^{14}\end{aligned}\right\}.$$

781. $$\left.\begin{aligned}\dot{x} &= 5x + y\cos y - \frac{x^3}{3}\\ \dot{y} &= 3x + 2y + \frac{x^4}{12} - y^3e^y\end{aligned}\right\}.$$

4. The Routh-Hurwitz criterion

Suppose we have a linear differential equation with constant real coefficients

$$a_0y^{(n)} + a_1y^{(n-1)} + \ldots + a_ny = 0 \tag{1}$$

$(a_0, a_1, \ldots, a_n = \text{const}, a_0 > 0)$.

The null solution $y \equiv 0$ of equation (1) is asymptotically stable if all the roots of the characteristic equation

$$f(\lambda) \equiv a_0\lambda^n + a_1\lambda^{n-1} + \ldots + a_n = 0 \tag{2}$$

have negative real parts.

The Routh-Hurwitz criterion. In order that all the roots of equation (2) should have negative real parts, it is necessary and sufficient that all the main diagonal minors of the Hurwitz matrix

$$\begin{pmatrix} a_1 & a_0 & 0 & 0 & 0 & 0 & \ldots & 0 \\ a_3 & a_2 & a_1 & a_0 & 0 & 0 & \ldots & 0 \\ a_5 & a_4 & a_3 & a_2 & a_1 & a_0 & \ldots & 0 \\ \cdot & \cdot & \cdot & \cdot & \cdot & \cdot & \cdot & \cdot \\ 0 & 0 & 0 & 0 & 0 & 0 & \ldots & a_n \end{pmatrix} \tag{3}$$

should be positive.

The Hurwitz matrix is made up as follows. Down the main diagonal are the coefficients of the polynomial (2), starting with a_1 and ending with a_n. The columns consist alternately of the coefficients with odd suffixes and the coefficients with even suffixes, the suffix of a_0 counting as even. All the remaining elements of the matrix which correspond to coefficients with suffix greater than n or less than 0 are taken to be zero.

The main diagonal minors of the Hurwitz matrix have the form

$$\Delta_1 = a_1, \quad \Delta_2 = \begin{vmatrix} a_1 & a_0 \\ a_3 & a_2 \end{vmatrix}, \quad \Delta_3 = \begin{vmatrix} a_1 & a_0 & 0 \\ a_3 & a_2 & a_1 \\ a_5 & a_4 & a_3 \end{vmatrix}, \ldots$$

$$\Delta_n = \begin{vmatrix} a_1 & a_0 & 0 & \ldots & 0 \\ a_3 & a_2 & a_1 & \ldots & 0 \\ a_5 & a_4 & a_3 & \ldots & 0 \\ \cdot & \cdot & \cdot & \cdot & \cdot \\ 0 & 0 & 0 & \ldots & a_n \end{vmatrix}.$$

Thus the Hurwitz condition states: for the stability of the solution $y \equiv 0$ of equation (1) it is necessary and sufficient that the relations

$$\Delta_1 > 0, \quad \Delta_2 > 0, \ldots, \Delta_n > 0 \tag{4}$$

are satisfied. Since $\Delta_n = a_n \Delta_{n-1}$, the condition $\Delta_n > 0$ can be replaced by the condition $a_n > 0$.

Example. Investigate the stability of the null solution of the equation

$$y^{IV} + 5y''' + 13y'' + 19y' + 10y = 0, \tag{5}$$

Solution. We form the characteristic equation

$$f(\lambda) \equiv \lambda^4 + 5\lambda^3 + 13\lambda^2 + 19\lambda + 10 = 0.$$

Here $a_0 = 1, a_1 = 5, a_2 = 13, a_3 = 19, a_4 = 10$. We write down the Hurwitz diagonal minors

$$\Delta_4 = \begin{vmatrix} 5 & 1 & 0 & 0 \\ 19 & 13 & 5 & 1 \\ 0 & 10 & 19 & 13 \\ 0 & 0 & 0 & 10 \end{vmatrix} = 6{,}240 > 0.$$

$$\Delta_3 = \begin{vmatrix} 5 & 1 & 0 \\ 19 & 13 & 5 \\ 0 & 10 & 19 \end{vmatrix} = 624 > 0,$$

$$\Delta_2 = \begin{vmatrix} 5 & 1 \\ 19 & 13 \end{vmatrix} = 46 > 0, \quad \Delta_1 = 5 > 0.$$

We have obtained $\Delta_1 > 0, \Delta_2 > 0, \Delta_3 > 0, \Delta_4 > 0$. Consequently the trivial solution $y \equiv 0$ of equation (5) is asymptotically stable.

The calculation can be organized in this way. We first form the largest Hurwitz minor Δ_n. From this it is easy to write down all the smaller minors $\Delta_{n-1}, \ldots, \Delta_1$. Then we start to calculate Δ_1, Δ_2, etc. If we meet a negative minor, the solution is unstable and we need not proceed further.

Investigate the stability of the null solution of the following equations:

782. $y''' - 3y' + 2y = 0.$

783. $y^{IV} + 4y''' + 7y'' + 6y' + 2y = 0.$

784. $y''' + 5y'' + 9y' + 5y = 0.$

785. $y^{IV} - 2y''' + y'' + 2y' - 2y = 0.$

786. $y^{IV} + 7y''' + 17y'' + 17y' + 6y = 0.$

787. $y''' - 3y'' + 12y' - 10y = 0.$

788. $y^{IV} + 5y''' + 18y'' + 34y' + 20y = 0.$

789. $y^{IV} + 7y''' + 19y'' + 23y' + 10y = 0.$

790. $y^{IV} + 11y''' + 41y'' + 61y' + 30y = 0.$

791. $y^{V} + 3y^{IV} - 5y''' - 15y'' + 4y' + 12y = 0.$

792. $y^{V} + 7y^{IV} + 33y''' + 88y'' + 122y' + 60y = 0.$

For what values of α is the null solution of these equations stable:

793. $y''' + 2y'' + \alpha y' + 3y = 0.$

794. $y^{IV} + \alpha y''' + 2y'' + y' + 3y = 0.$

795. $y^{IV} + 2y''' + \alpha y'' + y' + y = 0.$

For what values of α, β is the null solution of these equations stable:

796. $y''' + \alpha y'' + \beta y' + y = 0.$

797. $y^{IV} + 3y''' + \alpha y'' + 2y' + \beta y = 0.$

5. Geometrical criterion for stability (the Mikhailov criterion)

Suppose we have a linear differential equation of the nth order with constant real coefficients

$$a_0 y^{(n)} + a_1 y^{(n-1)} + \ldots + a_n y = 0. \tag{1}$$

Its characteristic equation is

$$f(\lambda) \equiv a_0 \lambda^n + a_1 \lambda^{n-1} + \ldots + a_n = 0. \tag{2}$$

The Mikhailov criterion enables us to solve the question of the arrangement of the roots of the characteristic equation (2) in the complex plane and consequently to solve the question of the stability of the null solution of (1).

Putting $\lambda = i\omega$, we obtain

$$f(i\omega) = u(\omega) + iv(\omega), \tag{3}$$

where

$$\left.\begin{aligned} u(\omega) &= a_n - a_{n-2}\omega^2 + a_{n-4}\omega^4 - \ldots \\ v(\omega) &= a_{n-1}\omega - a_{n-3}\omega^3 + \ldots \end{aligned}\right\}. \tag{4}$$

The quantity $f(i\omega)$, for a given value of the parameter ω, can be represented as a vector in the complex u, v plane, with its starting point at the origin.

As ω varies in the interval $(-\infty, +\infty)$ the end point of this vector describes a certain curve, the so-called Mikhailov curve (Fig. 27).

Since the function $u(\omega)$ is even, the Mikhailov curve is symmetrical about the axis Ou, so it is sufficient to draw that half of the curve which corresponds to the variation of ω from 0 to $+\infty$.

If the polynomial $f(\lambda)$ of degree n has m roots with positive real part and $n - m$ with negative real part, the angle ϕ through which the vector $f(i\omega)$ turns as ω varies from 0 to $+\infty$ is $\frac{1}{2}(n - 2m)\pi$.

It is clear that, for the solution of equation (1) to be stable, it is necessary and sufficient that $m = 0$.

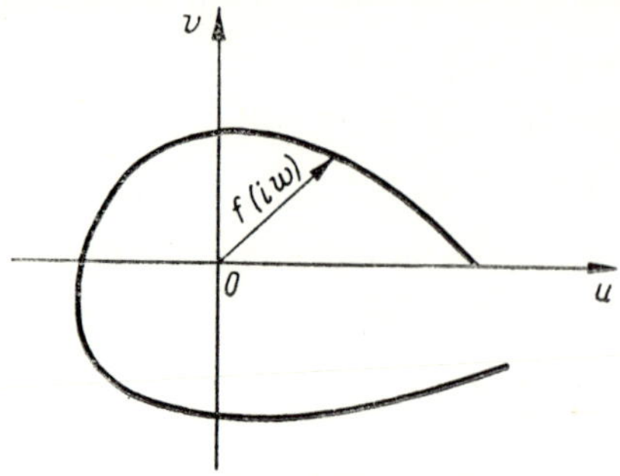

FIGURE 27

Mikhailov criterion. For the null solution $y \equiv 0$ of equation (1) to be stable, it is necessary and sufficient that:

(1) as ω varies from 0 to $+\infty$, the vector $f(i\omega)$ should turn through an angle $\phi = \frac{1}{2}n\pi$, i.e. it should make $\frac{1}{4}n$ revolutions counter-clockwise,

(2) as ω varies from 0 to $+\infty$, the hodograph of $f(i\omega)$ should not pass through the origin (0, 0).

Hence it follows that for the solution of equation (1) to be stable it is *necessary* that all the roots of the equations

$$u(\omega) = 0, \quad v(\omega) = 0$$

should be real and should alternate with each other, i.e. between any two roots of one equation there should lie a root of the other equation.

Example. Investigate the stability of the null solution $y \equiv 0$ of the equation

$$y^{IV} + y''' + 4y'' + y' + y = 0.$$

Solution. We form the characteristic polynomial

$$f(\lambda) = \lambda^4 + \lambda^3 + 4\lambda^2 + \lambda + 1.$$

Further,

$$f(i\omega) = \omega^4 - i\omega^3 - 4\omega^2 + i\omega + 1,$$

$$u(\omega) = \omega^4 - 4\omega^2 + 1,$$

$$v(\omega) = -\omega^3 + \omega = \omega(1 - \omega)(1 + \omega).$$

We draw the curve

$$\left.\begin{aligned} u &= u(\omega) \\ v &= v(\omega) \end{aligned}\right\}, \; 0 \leqslant \omega < +\infty$$

ω	0	$\sqrt{2-\sqrt{3}}$	1	$\sqrt{2+\sqrt{3}}$
u	1	0	-2	0
v	0	+	0	—

$$\lim_{\omega \to +\infty} \frac{v}{u} = 0$$

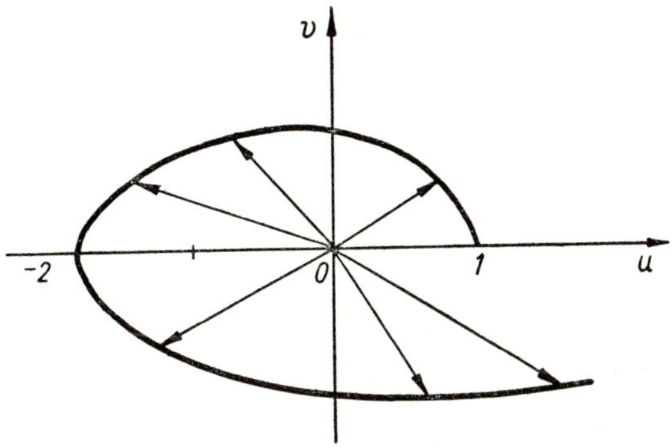

FIGURE 28

The angle of rotation of the radius vector is

$$\phi = 4 \, . \, \tfrac{1}{2}\pi = \tfrac{1}{2}(n - 2m)\pi.$$

Hence $n - 2m = 4$ and since $n = 4$, $m = 0$, i.e. all the roots of the characteristic equation lie in the left half-plane. This means that the trivial solution $y \equiv 0$ is asymptotically stable.

With the help of the Mikhailov criterion, investigate the stability of the null solution of the following equations:

798. $2y''' + 7y'' + 7y' + 2y = 0.$

799. $y''' + 2y'' + 2y' + y = 0.$

800. $2y^{IV} + 13y''' + 28y'' + 23y' + 6y = 0.$

801. $3y^{IV} + 13y''' + 19y'' + 11y' + 2y = 0.$

802. $2y^{IV} + 6y''' + 9y'' + 6y' + 2y = 0.$

803. $y^{IV} + 4y''' + 16y'' + 24y' + 20y = 0.$

804. $y^{V} + 13y^{IV} + 43y''' + 51y'' + 40y' + 12y = 0.$

805. $y''' + y = 0.$

806. $y^{IV} + y''' + y' + y = 0.$

807. $y^{V} + 3y^{IV} + 2y''' + y'' + 3y' + 2y = 0.$

808. $y^{V} + y^{IV} + y''' + y'' + y' + y = 0.$

809. $2y^{IV} + 11y''' + 21y'' + 16y' + 4y = 0.$

810. $y^{VI} + y^{V} + y^{IV} + y'' + y' + y = 0.$

811. $2y^{IV} + 9y''' + 32y'' + 54y' + 20y = 0.$

812. $6y^{IV} + 29y''' + 45y'' + 24y' + 4y = 0.$

813. $y^{V} + y^{IV} + 2y''' + 2y'' + 2y' + 2y = 0.$

814. $y^{VI} + y^{V} + 3y^{IV} + 2y''' + 4y'' + 2y' + 2y = 0.$

815. $y^{V} + 2y^{IV} + y''' + 2y'' + y' + 2y = 0.$

§18. AN OPERATIONAL METHOD AND ITS APPLICATION TO THE SOLUTION OF DIFFERENTIAL EQUATIONS

1. The Laplace transformation and its main properties

Object function and transform.

An *object function* is a complex-valued function of a real variable, $f(t)$, which satisfies the following properties:

(1) $f(t) = 0$ if $t < 0$;

(2) $f(t)$ and its derivatives to sufficiently high order are continuous on the whole of the t axis, except for isolated points, at which $f(t)$ or its derivatives have a discontinuity of the first kind, where on each finite interval of the t axis there are only a finite number of such points;

(3) as t increases the modulus of the function $f(t)$ increases no faster than a certain exponential function, i.e. there exist constants $M > 0$ and $s_0 \geqslant 0$ such that, for all t,

$$|f(t)| < Me^{s_0 t}. \tag{1}$$

The number s_0 is called the *growth index* of the function $f(t)$.

The *transform* of the object function (according to Laplace) is a function of a complex variable $p = s + i\sigma$, defined by the formula

$$F(p) = \int_0^{\infty} f(t)e^{-pt}\,dt \tag{2}$$

if $\operatorname{Re} p > s_0$, where s_0 is the growth index of $f(t)$.

The transformation (2), which relates the object function $f(t)$ to its transform $F(p)$ is called a Laplace transformation. We write

$$f(t) \risingdotseq F(p).$$

The following **theorem** holds:

If $f(t) \risingdotseq F(p)$, then at any point at which it is continuous the function $f(t)$ is defined as

$$f(t) = \frac{1}{2\pi i} \int_{a-i\infty}^{a+i\infty} e^{pt} F(p)\, dp, \tag{3}$$

where

$$\int_{a-i\infty}^{a+i\infty} e^{pt} F(p)\, dp = \lim_{b \to +\infty} \int_{a-ib}^{a+ib} e^{pt} F(p)\, dp$$

(The formula (3) is called the inversion formula for the Laplace transformation.)

Properties of the Laplace transformation

I. Property of linearity. For any complex constants α and β

$$\alpha f(t) + \beta g(t) \risingdotseq \alpha F(p) + \beta G(p) \tag{4}$$

(here, and in the later work, we suppose that $f(t) \risingdotseq F(p)$, $g(t) \risingdotseq G(p)$).

II. Theorem of similarity. For any constant $\alpha > 0$,

$$f(\alpha t) \risingdotseq \frac{1}{\alpha} F\left(\frac{p}{\alpha}\right). \tag{5}$$

III. Differentiation of the object function. If $f'(t)$ is the object function,

$$f'(t) \risingdotseq pF(p) - f(0) \tag{6}$$

Generalization. If $f(t)$ is continuously differentiable n times on $(0, +\infty)$, and if $f^{(n)}(t)$ is the object function, then

$$f^{(n)}(t) \risingdotseq p^n F(p) - p^{n-1} f(0) - p^{n-2} f'(0) - \ldots - f^{(n-1)}(0). \tag{7}$$

IV. Differentiation of the transform is equivalent to multiplication of the object function by minus the argument, i.e.

$$F'(p) \risingdotseq -t \,.\, f(t). \tag{8}$$

Generalization:

$$F^{(n)}(p) \risingdotseq (-1)^n t^n f(t). \tag{9}$$

V. Integration of the object function reduces to division of the transform by p:

$$\int_0^t f(t)\,dt \risingdotseq \frac{F(p)}{p}. \tag{10}$$

VI. Integration of the transform is equivalent to division of the object function by t:

$$\int_p^\infty F(p)\,dp \risingdotseq \frac{f(t)}{t} \tag{11}$$

(we assume that the integral $\int_p^\infty F(p)\,dp$ converges).

VII. Delay theorem. For any positive number τ,

$$f(t-\tau) \risingdotseq e^{-p\tau}F(p). \tag{12}$$

VIII. Shifting theorem (multiplication of the object function by an exponential function). For any complex number λ,

$$e^{\lambda t}f(t) \risingdotseq F(p-\lambda). \tag{13}$$

IX. Multiplication theorem (E. Borel). The product of two transforms $F(p)$ and $G(p)$ is also a transform, where

$$F(p)G(p) \risingdotseq \int_0^t f(\tau)g(t-\tau)\,d\tau. \tag{14}$$

The integral on the right hand side of (14) is called the *convolution* function of $f(t)$ and $g(t)$ and is denoted by the symbol

$$(f_* g) = \int_0^t f(\tau)g(t-\tau)\,d\tau.$$

So theorem IX states that the *multiplication of transforms is equivalent to convolution of the object functions*:

$$F(p)G(p) \risingdotseq (f_* g). \tag{15}$$

X. Theorem of rationality of the transform. In order that the transform $F(p)$ should be a rational function, it is necessary and sufficient that the object function $f(t)$ should be a linear combination of functions of the form $t^m e^{\lambda t}$ (where m is a non-negative integer and λ is complex).

XI. Calculation of the object function of a rational transform. Suppose $F(p)$ is a proper rational fraction, whose expansion in partial fractions is

$$F(p) = \sum_k \sum_{r=1}^{n_k} \frac{M_{kr}}{(p - p_k)^r}, \tag{16}$$

where M_{kr} and p_k are certain complex numbers. Then

$$f(t) = \sum_k \sum_{r=1}^{n_k} \frac{M_{kr} t^{r-1}}{(r-1)!} e^{p_k t} \tag{17}$$

is the object function which has the transform $F(p)$. In particular, if all the poles of $F(p)$ are simple, then

$$f(t) = \sum_k M_k e^{p_k t}, \tag{18}$$

where $M_k = \operatorname*{Res}_{p_k} F(p)$.

Generalization. If the function $F(p) = \dfrac{A(p)}{B(p)}$ is rational, where the degree of the polynomial $A(p)$ is less than the degree of the polynomial $B(p)$, then the object function of $F(p)$ is the function

$$f(t) = \sum_k \frac{1}{(n_k - 1)!} \lim_{p \to p_k} \frac{d^{n_k - 1}}{dp^{n_k - 1}} \{F(p)(p - p_k)^{n_k} e^{pt}\}, \tag{19}$$

where p_k are the poles of $F(p)$, n_k their multiplicities, and the sum is taken over all the poles.

In particular, if all the poles of $F(p)$ are simple, then formula (19) can be simplified to the form

$$f(t) = \sum_k \frac{A(p_k)}{B'(p_k)} e^{p_k t}. \tag{20}$$

Examples.

1. Find the transform of the Heaviside unit function $\eta(t)$ (Fig. 29):

$$\eta(t) = \begin{cases} 0, & \text{if } t < 0 \\ 1, & \text{if } t > 0. \end{cases} \tag{21}$$

Solution. According to (2) and (21) we have:

$$\eta(t) \risingdotseq \int_0^\infty e^{-pt} \eta(t)\, dt = \int_0^\infty e^{-pt}\, dt = \frac{1}{p} \; (\operatorname{Re} p > 0).$$

So

$$\eta(t) \risingdotseq \frac{1}{p}. \tag{22}$$

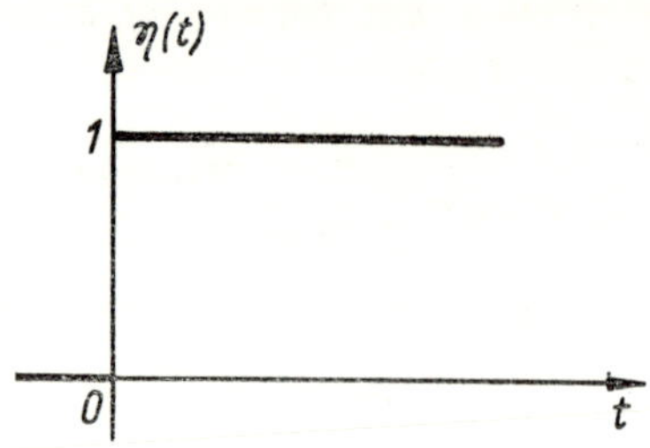

FIGURE 29

2. Find the transform of the step-function (Fig. 30).

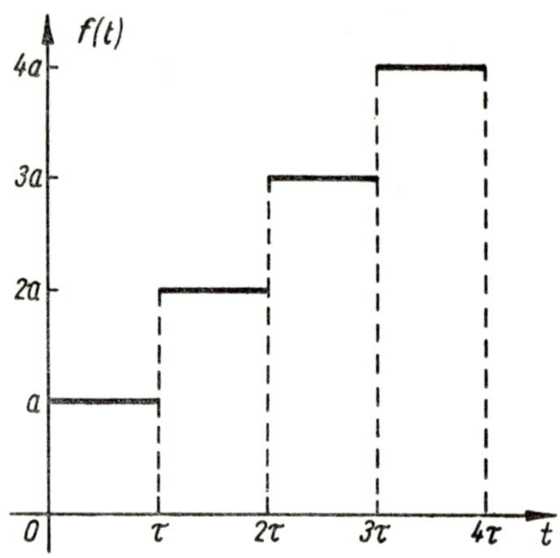

FIGURE 30

Solution. We have $f(t) = a\{\eta(t) + \eta(t-\tau) + \eta(t-2\tau) + \ldots\}$. From the delay theorem we obtain

$$f(t) \doteqdot a\left(\frac{1}{p} + \frac{1}{p}e^{-p\tau} + \frac{1}{p}e^{-2p\tau} + \ldots\right).$$

In the brackets we have a geometrical progression whose ratio is $q = e^{-p\tau}$. Since $|q| = |e^{-p\tau}| = e^{-s\tau} < 1$, the geometrical progression converges, and we obtain

$$f(t) \doteqdot \frac{a}{p}\frac{1}{1-e^{-p\tau}}.$$

3. Find the transform of the function $f(t) = e^{\lambda t}$, where λ is an arbitrary complex number.

Solution.

$$e^{\lambda t} \risingdotseq \int_0^\infty e^{-pt} e^{\lambda t}\,dt = \int_0^\infty e^{-(p-\lambda)t}\,dt = \frac{1}{p-\lambda}.$$

So

$$e^{\lambda t} \risingdotseq \frac{1}{p-\lambda}. \tag{23}$$

In particular, if $\lambda = 1$ we obtain

$$e^t \risingdotseq \frac{1}{p-1},$$

and if $\lambda = -1$

$$e^{-t} \risingdotseq \frac{1}{p+1}.$$

4. Find the transform of the function $f(t) = \operatorname{ch} t$.

Solution. We have $\operatorname{ch} t = \frac{1}{2}(e^t + e^{-t})$. Using the property of linearity and the results of the previous example, we obtain

$$\operatorname{ch} t \risingdotseq \frac{1}{2}\left(\frac{1}{p-1} + \frac{1}{p+1}\right) = \frac{p}{p^2-1}. \tag{24}$$

5. Find the transform of the function $f(t) = \sin \alpha t \cos \beta t$.

Solution. Since $\sin \alpha t + \cos \beta t = \frac{1}{2}[\sin(\alpha+\beta)t + \sin(\alpha-\beta)t]$, the problem reduces to the problem of finding the transform of the function $\sin \omega t$. We have

$$\sin \omega t = \frac{1}{2i}(e^{\omega t i} - e^{-\omega t i}) \risingdotseq \frac{1}{2i}\left(\frac{1}{p-\omega i} - \frac{1}{p+\omega i}\right) = \frac{\omega}{p^2+\omega^2}, \tag{25}$$

and then

$$\sin \alpha t \cos \beta t \risingdotseq \frac{1}{2}\left[\frac{\alpha+\beta}{p^2+(\alpha+\beta)^2} + \frac{\alpha-\beta}{p^2+(\alpha-\beta)^2}\right];$$

after simple transformation we obtain finally

$$\sin \alpha t \cos \beta t \risingdotseq \frac{\alpha(p^2+\alpha^2-\beta^2)}{(\alpha^2+\beta^2+p^2)^2 - 4a^2\beta^2}.$$

6. Find the object function of the given transform

$$F(p) = \frac{1}{(p^2+1)^2}.$$

Solution. First method. Using the multiplication theorem and the fact that $\frac{1}{p^2+1} \risingdotseq \sin t$, we have

$$\frac{1}{(p^2+1)^2} = \frac{1}{p^2+1} \cdot \frac{1}{p^2+1} \risingdotseq \int_0^t \sin \tau \sin (t-\tau)\, d\tau$$

$$= \frac{1}{2}\int_0^t [\cos (2\tau - t) - \cos t]\, d\tau = \frac{1}{2}(\sin t - t \cos t).$$

Second method. We know that $\sin t \risingdotseq \frac{1}{p^2+1}$. From formula (8) for differentiating the transform, we obtain

$$-t \sin t \risingdotseq \left(\frac{1}{p^2+1}\right)' = -\frac{2p}{(p^2+1)^2}.$$

Hence, using the formula (10) for integrating the object function, we obtain finally

$$\frac{1}{(p^2+1)^2} \risingdotseq \frac{1}{2}\int_0^t t \sin t\, dt = \frac{1}{2}(\sin t - t \cos t).$$

Third method. From formula (19) we have

$$f(t) = \frac{1}{2\pi i}\int_L \frac{e^{pt}\, dp}{(p^2+1)^2} = \sum_{p_k} \text{Res}\, \frac{e^{p_k t}}{(p_k^2+1)^2},$$

where the sum is taken over all the poles of the function $\frac{e^{pt}}{(p^2+1)^2}$, having regard to their multiplicities, and L is a closed contour containing the poles.

In our case the poles are the complex numbers $p_1 = i$, $p_2 = -i$, both of multiplicity 2, i.e. $n_1 = 2$ and $n_2 = 2$, so (19) gives

$$(a)\ \text{Res}\, \frac{e^{p_1 t}}{(p_1^2+1)^2} = \lim_{p\to i} \frac{d}{dp}\left[\frac{(p-i)^2 e^{pt}}{(p^2+1)^2}\right] = \lim_{p\to i} \frac{d}{dp}\left[\frac{e^{pt}}{(p+i)^2}\right]$$

$$= \lim_{p\to i} \frac{(p+i)^2 t e^{pt} - e^{pt} 2(p+i)}{(p+i)^4}$$

$$= \lim_{p\to i} \frac{(p+i)t - 2}{(p+i)^3}\, e^{pt} = \frac{1-it}{4i}\, e^{it},$$

(*b*) $\operatorname{Res} \dfrac{e^{p_2 t}}{(p_2^2+1)^2} = -\dfrac{1+it}{4i} e^{-it},$

thus

$$f(t) = \operatorname{Res} \frac{e^{p_1 t}}{(p_1^2+1)^2} + \operatorname{Res} \frac{e^{p_2 t}}{(p_2^2+1)^2} = \frac{1}{2}(\sin t - t \cos t).$$

7. Find the object function of the given transform

$$F(p) = \frac{1}{(p-1)^3(p^3+1)}.$$

Solution. The expression for $F(p)$ in partial fractions is

$$F(p) = \frac{3}{8}\frac{1}{p-1} - \frac{3}{4}\frac{1}{(p-1)^2} + \frac{1}{2}\frac{1}{(p-1)^3} - \frac{1}{24}\frac{1}{p+1}$$
$$-\frac{1}{3}\frac{p-2}{p^2-p+1}.$$

Finding the object function for each term, we obtain

$$f(t) = \frac{3}{8}e^t - \frac{3}{4}te^t + \frac{1}{4}t^2e^t - \frac{1}{24}e^{-t} - \frac{1}{3}e^{\frac{t}{2}}\cos\frac{\sqrt{3}}{2}t$$
$$+\frac{\sqrt{3}}{3}e^{\frac{t}{2}}\sin\frac{\sqrt{3}}{2}t.$$

8. Find the object function of the given transform

$$F(p) = \frac{2p+3}{p^3+4p^2+3p}.$$

Solution. First method. Expressing $F(p)$ in partial fractions, we obtain

$$F(p) = \frac{1}{p} - \frac{1}{2}\cdot\frac{1}{p+1} - \frac{1}{2}\cdot\frac{1}{p+3}.$$

Finding the object function for each term, we obtain

$$f(t) = 1 - \tfrac{1}{2}e^{\ t} - \tfrac{1}{2}c^{-3t}.$$

Second method. We use formula (20). Since the poles of $F(p)$ are

$$p_1 = 0, \quad p_2 = -1, \quad p_3 = -3,$$

$$A(p) = 2p+3, \quad B(p) = p^3+4p^2+3p, \quad B'(p) = 3p^2+8p+3,$$

we have

$$f(t) = \frac{A(p_1)}{B'(p_1)} e^{p_1 t} + \frac{A(p_2)}{B'(p_2)} e^{p_2 t} + \frac{A(p_3)}{B'(p_3)} e^{p_3 t} = 1 - \frac{1}{2} e^{-t} - \frac{1}{2} e^{-3t}.$$

In the following examples, find the transforms of the given object functions:

816. $t^2 - 2t + 2.$

817. $t^3 + 4t^2 + 4t.$

818. $(t-2)^3\eta(t-2).$

819. $te^{-\alpha t}.$

820. $(t+2)te^t.$

821. $\mathrm{ch}^2\, at.$

822. $(t-1)^2e^{1-t}.$

823. $e^{\alpha t} \sin \beta t.$

824. $e^{3t} \cos 3t \cos 4t.$

825. $e^{\lambda(t-\alpha)} \sin (t-\alpha)\eta(t-\alpha).$

826. $e^{2t} \sin \left(t + \frac{\pi}{4}\right).$

827. $e^{\alpha t} \cos (t + \beta).$

828. $\frac{\sin t}{t}.$

829. $e^{-\lambda t} \frac{\sin t}{t}.$

830. $\sin 5t \sin 2t.$

831. $\sin^2 2t.$

832. $e^{(-2+i)t}.$

833. $e^{(\alpha+i\beta)t}.$

834. $\cos 2t \cos 4t.$

835. $\cos^2 4t.$

In the following examples, find the object functions of the given transforms:

836. $\frac{2p+3}{p^3+4p^2+5p}.$

837. $\frac{p^2+a^2}{(p^2-a^2)^2}$ (a – constant).

838. $\frac{k!}{p^{k+1}}.$

839. $\frac{2}{(p-1)(p-3)}.$

840. $\frac{3p+19}{2p^2+8p+19}.$

841. $\frac{1}{(p^2+p+1)^2}.$

842. $\dfrac{1}{(p-1)^2(p+2)}$.

843. $\dfrac{2p^2-2\sqrt{2}p}{p^4-3p^2+2}$.

844. $\dfrac{1}{p^2+p+1}$.

845. $\dfrac{1}{p^4-1}$.

846. $\dfrac{pe^{-2p}}{p^2+4}$.

847. $\dfrac{e^{-\frac{p}{2}}}{p^2+9}$.

848. $\dfrac{p^3+9p^2+27p+25}{(p+1)^3(p+2)^2}$.

849. $\dfrac{2p+5}{p^2-6p+12}$.

850. $\dfrac{1}{p^4+2p^2-3}$.

851. $\dfrac{e^{-\frac{3}{2}p}}{p^2}$.

2. Linear equations with constant coefficients

Suppose we are given a differential equation of the second order with constant coefficients

$$x''(t) + a_1x'(t) + a_2x(t) = f(t) \tag{1}$$

and the initial conditions

$$x(0) = x_0, \quad x'(0) = x_1. \tag{2}$$

We suppose that the function $f(t)$ and the solution $x(t)$, together with their derivatives up to the second order, are object functions. We denote

$$x(t) \risingdotseq X(p), \quad f(t) \risingdotseq F(p).$$

For equation (1) and the initial conditions (2) the operator equation has the form

$$(p^2 + a_1 p + a_2)X(p) = F(p) + x_0(p + a_1) + x_1. \tag{3}$$

Solving equation (3), we find the operator solution

$$X(p) = \frac{F(p) + x_0(p + a_1) + x_1}{p^2 + a_1 p + a_2}. \tag{4}$$

Finding the object function of $X(p)$, we obtain the solution of equation (1) which satisfies the initial conditions (2).

Similarly we can solve any equation of the nth order with constant coefficients and initial conditions for $t = 0$.

Example. Solve the equation

$$x'' - 5x' + 4x = 4, \quad x(0) = 0, \quad x'(0) = 2.$$

Solution. Since $4 \risingdotseq \frac{4}{p}$ and $x_0 = x(0) = 0$, $x_1 = x'(0) = 2$, the operator equation has the form

$$(p^2 - 5p + 4)X(p) = \frac{4}{p} + 2.$$

Hence we find the operator solution

$$X(p) = \frac{2p + 4}{p(p^2 - 5p + 4)}.$$

We express the right hand side in partial fractions:

$$X(p) = \frac{1}{p} - \frac{2}{p - 1} + \frac{1}{p - 4}.$$

Going over to the object functions, we obtain the required solution

$$x(t) = 1 - 2e^t + e^{4t}.$$

Example. Solve the equation

$$x'' + 4x' + 4x = 8e^{-2t}, \quad x(0) = 1, \quad x'(0) = 1.$$

Solution. Since $8e^{-2t} \risingdotseq \frac{8}{p + 2}$ and $x_0 = x_1 = 1$, the operator equation has the form

$$(p^2 + 4p + 4)X(p) = \frac{8}{p + 2} + p + 4 + 1,$$

and consequently the operator solution is

$$X(p) = \frac{p^2 + 7p + 18}{(p + 2)^3}.$$

Expressing the right hand side in partial fractions, we have

$$X(p) = \frac{8}{(p + 2)^3} + \frac{3}{(p + 2)^2} + \frac{1}{p + 2}.$$

Going over to the object functions, we obtain the solution of the given problem:

$$x(t) = 4t^2e^{-2t} + 3te^{-2t} + e^{-2t}.$$

Solve the following equations:

852. $x' + 3x = e^{-2t}, \quad x(0) = 0.$

853. $x' - 3x = 3t^3 + 3t^2 + 2t + 1, \quad x(0) = -1.$

854. $x' - x = \cos t - \sin t, \quad x(0) = 0.$

855. $x' + x = 2 \sin t, \quad x(0) = 0.$

856. $2x' + 6x = te^{-3t}, \quad x(0) = -\tfrac{1}{2}.$

857. $x'' + 4x' + 3x = 1, \quad x(0) = 3, \quad x'(0) = -2.$

858. $x'' - 2x' + 2x = 1, \quad x(0) = \tfrac{1}{2}, \quad x'(0) = 0.$

859. $x'' - 5x' + 6x = 12, \quad x(0) = 2, \quad x'(0) = 0.$

860. $x'' + 3x' - 1 = 0, \quad x(0) = 0, \quad x'(0) = \tfrac{1}{3}.$

861. $x'' - 2x' + 1 = 0, \quad x(0) = x'(0) = \tfrac{1}{2}.$

862. $x'' + 3x' + 2x = 2t^2 + 1, \quad x(0) = 4, \quad x'(0) = -3.$

863. $x'' - 2x' - 3x = 3 + 7t + 3t^2, \quad x(0) = x'(0) = -1.$

864. $x'' - 7x' = -(14t + 5), \quad x(0) = 2, \quad x'(0) = 8.$

865. $x'' + 2x' = 6t^2 + 6t - 3, \quad x(0) = 0, \quad x'(0) = -\tfrac{3}{2}.$

866. $x'' + 6x' = t, \quad x(0) = 0, \quad x'(0) = -\tfrac{1}{36}.$

867. $x'' + x = 2e^t, \quad x(0) = 1, \quad x'(0) = 2.$

868. $7x'' + 14x' = (t - \tfrac{1}{4})e^{-2t}, \quad x(0) = 2, \quad x'(0) = -\tfrac{1}{56}.$

869. $x'' - 4x' + 4x = (t - 1)e^{2t}, \quad x(0) = 0, \quad x'(0) = 1.$

870. $4x'' - 4x' + x = e^{\frac{t}{2}}, \quad x(0) = -2, \quad x'(0) = 0.$

871. $x'' + 3x' + 2x = e^{-t} + e^{-2t}, \quad x(0) = 2, \quad x'(0) = -3.$

872. $x'' - x' - 6x = 6e^{3t} + 2e^{-2t}, \quad x(0) = 0, \quad x'(0) = \tfrac{4}{5}.$

873. $x'' + 4x' + 4x = t^2e^{-2t}, \quad x(0) = x'(0) = 0.$

874. $x'' - x' = 2 \sin t, \quad x(0) = 2, \quad x'(0) = 0.$

875. $x'' + 9x = 18 \cos 3t, \quad x(0) = 0, \quad x'(0) = 9.$

876. $x'' + 4x = 4 \cos 2t - \frac{1}{2} \sin 2t, \quad x(0) = 0, \quad x'(0) = \frac{1}{8}.$

877. $x'' + 2x' + 3x = t \cos t, \quad x(0) = -\frac{1}{4}, \quad x'(0) = 0.$

878. $x'' - 2x' + 10x = \cos 3t, \quad x(0) = 1, \quad x'(0) = \frac{18}{37}.$

879. $x'' - 4x' + 5x = 2e^{2t}(\sin t + \cos t), \quad x(0) = 1, \quad x'(0) = 2.$

880. $x''' - x'' = 0, \quad x(0) = 1, \quad x'(0) = 3, \quad x''(0) = 2.$

881. $x''' - x = 2, \quad x(0) = 0, \quad x'(0) = \frac{1}{2}(1 + \sqrt{3}), \quad x''(0) = \frac{1}{2}(1 - \sqrt{3}).$

882. $x''' + x'' - 2x = te^t, \quad x(0) = 2, \quad x'(0) = \frac{21}{25}, \quad x''(0) = -\frac{28}{25}.$

883. $x'' + 4x' = 8\sqrt{2} \sin \left(t + \frac{\pi}{4}\right), \quad x(0) = -\frac{40}{17}, \quad x'(0) = \frac{24}{17}.$

884. $x'' + 4x = 8\sqrt{2} \sin \left(2t + \frac{\pi}{4}\right), \quad x(0) = 0, \quad x'(0) = -2.$

3. Systems of linear differential equations

Suppose we have to find the solution of a system of two equations with constant coefficients

$$\left.\begin{aligned} \frac{dx}{dt} &= a_1 x + b_1 y + f_1(t) \\ \frac{dy}{dt} &= a_2 x + b_2 y + f_2(t) \end{aligned}\right\}, \tag{1}$$

which satisfies the initial conditions

$$x(0) = x_0, \quad y(0) = y_0. \tag{2}$$

We consider the transforms of the unknown functions, their derivatives and the functions $f_1(t), f_2(t)$:

$$\begin{aligned} x(t) &\fallingdotseq X(p), & y(t) &\fallingdotseq Y(p), \\ x'(t) &\fallingdotseq pX(p) - x_0, & y'(t) &\fallingdotseq pY(p) - y_0, \\ f_1(t) &\fallingdotseq F_1(p), & f_2(t) &\fallingdotseq F_2(p) \end{aligned}$$

and form the operator system

$$\left.\begin{aligned} pX(p) &= a_1 X(p) + b_1 Y(p) + F_1(p) + x_0 \\ pY(p) &= a_2 X(p) + b_2 Y(p) + F_2(p) + y_0 \end{aligned}\right\}. \tag{3}$$

This is a linear system of two equations with two unknowns $X(p)$ and $Y(p)$. Solving it, we find $X(p)$ and $Y(p)$ and then, going over to the object functions, we obtain the solution $x(t)$, $y(t)$ of the system (1) which satisfies the initial conditions (2). Similarly we can solve linear systems of the form

$$\frac{dx_k}{dt} = \sum_{l=1}^{n} a_{kl}x_l, \; a_{kl} = \text{const}; \quad x_k(0) = x_k^0. \quad (k = 1, 2, \ldots, n)$$

Example. Find the solution of the system

$$\left.\begin{aligned} \frac{dx}{dt} &= -7x + y + 5 \\ \frac{dy}{dt} &= -2x - 5y - 37t \end{aligned}\right\},$$

which satisfies the initial conditions $x(0) = 0$, $y(0) = 0$.

Solution. Since $5 \risingdotseq \frac{5}{p}$, $-37t \risingdotseq -\frac{37}{p^2}$ and $x_0 = y_0 = 0$, the operator system has the form:

$$\left.\begin{aligned} pX(p) &= -7X(p) + Y(p) + \frac{5}{p} \\ pY(p) &= -2X(p) - 5Y(p) - \frac{37}{p^2} \end{aligned}\right\}.$$

Solving it, we obtain

$$X(p) = \frac{5p^2 + 25p - 37}{p^2(p^2 + 12p + 37)}, \quad Y(p) = \frac{-47p - 259}{p^2(p^2 + 12p + 37)}.$$

We express the right hand sides in partial fractions:

$$X(p) = \frac{1}{p} - \frac{1}{p^2} - \frac{p+6}{p^2 + 12p + 37},$$

$$Y(p) = \frac{1}{p} - \frac{7}{p^2} - \frac{p+5}{p^2 + 12p + 37}$$

or

$$X(p) = \frac{1}{p} - \frac{1}{p^2} - \frac{p+6}{(p+6)^2 + 1},$$

$$Y(p) = \frac{1}{p} - \frac{7}{p^2} - \frac{p+6}{(p+6)^2 + 1} + \frac{1}{(p+6)^2 + 1}.$$

Going over to the object functions, we obtain the required solution

$$\left.\begin{aligned} x(t) &= 1 - t - e^{-6t}\cos t \\ y(t) &= 1 - 7t - e^{-6t}\cos t + e^{-6t}\sin t \end{aligned}\right\}.$$

In the following problems, solve the systems of equations by the operational method:

885. $\left.\begin{aligned} \frac{dx}{dt} + y = 0 \\ \frac{dy}{dt} + x = 0 \end{aligned}\right\}, \quad x(0) = 2, \quad y(0) = 0.$

886. $\left.\begin{aligned} \frac{dx}{dt} + x - 2y = 0 \\ \frac{dy}{dt} + x + 4y = 0 \end{aligned}\right\}, \quad x(0) = y(0) = 1.$

887. $\left.\begin{aligned} \frac{dx}{dt} &= -y \\ \frac{dy}{dt} &= 2(x + y) \end{aligned}\right\}, \quad x(0) = y(0) = 1.$

888. $\left.\begin{aligned} \frac{dx}{dt} + 2y = 3t \\ \frac{dy}{dt} - 2x = 4 \end{aligned}\right\}, \quad x(0) = 2, \quad y(0) = 3.$

889. $\left.\begin{aligned} \frac{dx}{dt} + x = y + e^t \\ \frac{dy}{dt} + y = x + e^t \end{aligned}\right\}, \quad x(0) = y(0) = 1.$

890. $\left.\begin{aligned} \frac{dx}{dt} + \frac{dy}{dt} = y + e^t \\ 2\frac{dx}{dt} + \frac{dy}{dt} + 2y = \cos t \end{aligned}\right\}, \quad x(0) = y(0) = 0.$

891. $\left.\begin{aligned} \frac{dx}{dt} &= y - z \\ \frac{dy}{dt} &= x + y \\ \frac{dz}{dt} &= x + z \end{aligned}\right\}, \quad x(0) = 1, \quad y(0) = 2, \quad z(0) = 3.$

892. $\left.\begin{aligned} \frac{dx}{dt} &= 4y + z \\ \frac{dy}{dt} &= z \\ \frac{dz}{dt} &= 4y \end{aligned}\right\}, \quad x(0) = 5, \quad y(0) = 0, \quad z(0) = 4.$

893. $\left.\begin{aligned} &\frac{dx}{dt} + 2\frac{dy}{dt} + x + y + z = 0 \\ &\frac{dx}{dt} + \frac{dy}{dt} + x + z = 0 \\ &\frac{dz}{dt} + 2\frac{dy}{dt} - y = 0 \end{aligned}\right\}, \quad x(0) = y(0) = 1, \quad z(0) = -4.$

894. $\left.\begin{aligned} &\frac{dx}{dt} - \frac{dy}{dt} - 2x + 2y = 1 - 2t \\ &\frac{d^2x}{dt^2} + 2\frac{dy}{dt} + x = 0 \end{aligned}\right\}, \quad x(0) = y(0) = x'(0) = 0.$

895. $\left.\begin{aligned} \frac{d^2x}{dt^2} &= y \\ \frac{d^2y}{dt^2} &= x \end{aligned}\right\}, \quad x(0) = y(0) = 1, \quad x'(0) = 2, \quad y'(0) = 0.$

896. $\left.\begin{aligned} \frac{d^2x}{dt^2} &= x - 4y \\ \frac{d^2y}{dt^2} &= -x + y \end{aligned}\right\}, \quad \begin{aligned} &x(0) = 2, \quad y(0) = 0, \\ &x'(0) = -\sqrt{3}, \quad y'(0) = \frac{\sqrt{3}}{2}. \end{aligned}$

897. $$\left.\begin{aligned}\frac{d^2x}{dt^2}+\frac{dy}{dt}&=e^t-x\\ \frac{d^2y}{dt^2}+\frac{dx}{dt}&=1\end{aligned}\right\},\quad \begin{aligned}&x(0)=1,\quad y(0)=0,\\ &x'(0)=2,\quad y'(0)=-1.\end{aligned}$$

898. $$\left.\begin{aligned}\frac{d^2x}{dt^2}+x+y&=5\\ \frac{d^2y}{dt^2}-4x-3y&=-3\end{aligned}\right\},\quad x(0)=y(0)=x'(0)=y'(0)=0.$$

899. $$\left.\begin{aligned}\frac{dx}{dt}+4y+2x&=4t+1\\ \frac{dy}{dt}+x-y&=\frac{3}{2}t^2\end{aligned}\right\},\quad x(0)=y(0)=0.$$

900. $$\left.\begin{aligned}\frac{dx}{dt}+y-2x&=0\\ \frac{dy}{dt}+x-2y&=-5e^t\sin t\end{aligned}\right\},\quad x(0)=2,\quad y(0)=3.$$

BIBLIOGRAPHY

1. R. L. Goodstein, *Mathematical Analysis* (Oxford University Press, 1948)
2. E. L. Ince, *Ordinary Differential Equations* (Longmans, Green, 1927)
3. W. Leighton, *Ordinary Differential Equations* (Longmans, 1964)
4. W. T. Martin and E. Reissner, *Elementary Differential Equations* Addison Wesley, 1961)
5. R. A. Struble, *Non-linear Differential Equations* (McGraw Hill, 1962)
6. L. S. Pontryagin, *Ordinary Differential Equations* (Addison Wesley and Pergamon, 1962)
7. M. Tenenbaum and H. Pollard, *Ordinary Differential Equations* (Harper and Row, 1963)

ANSWERS TO PROBLEMS

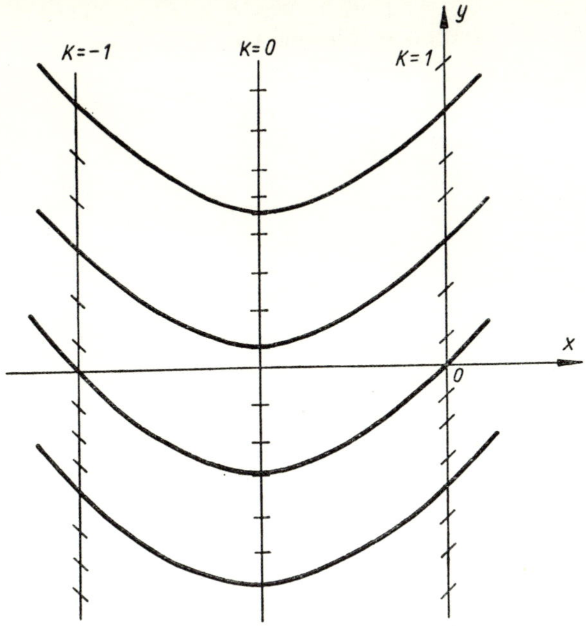

Answer to Problem 40

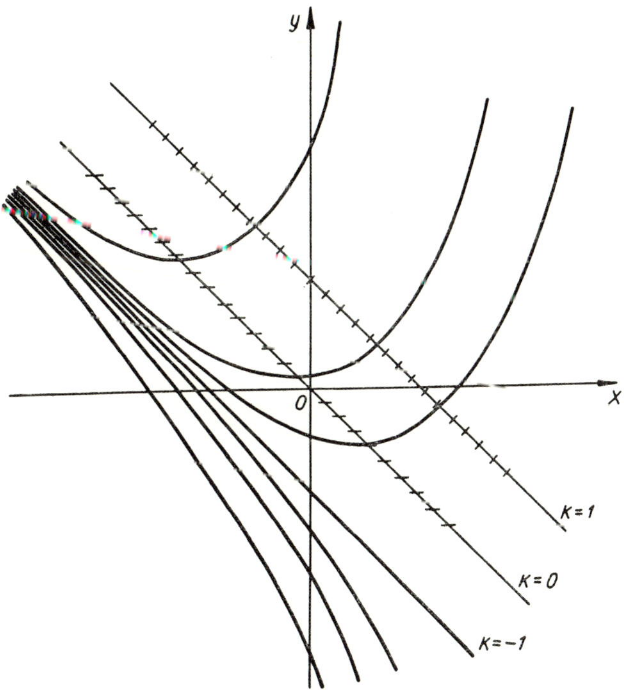

Answer to Problem 41

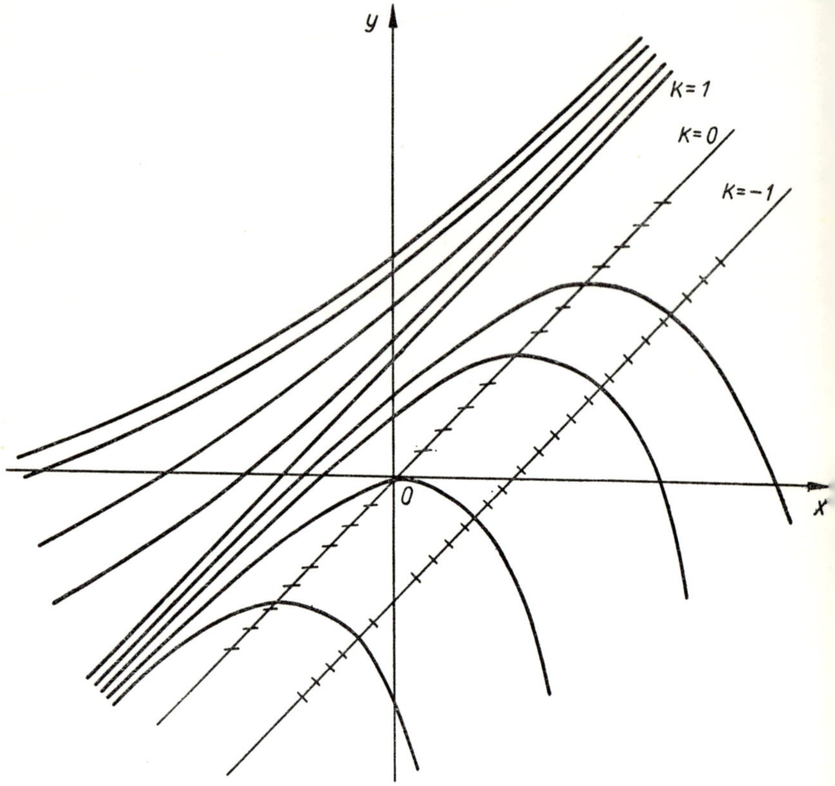

Answer to Problem 42

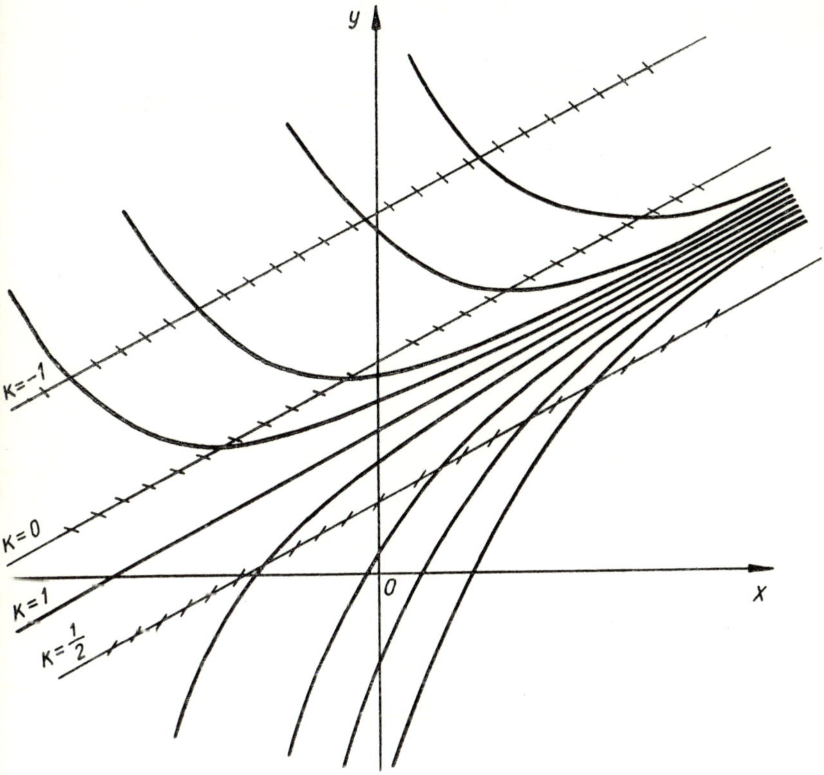

Answer to Problem 43

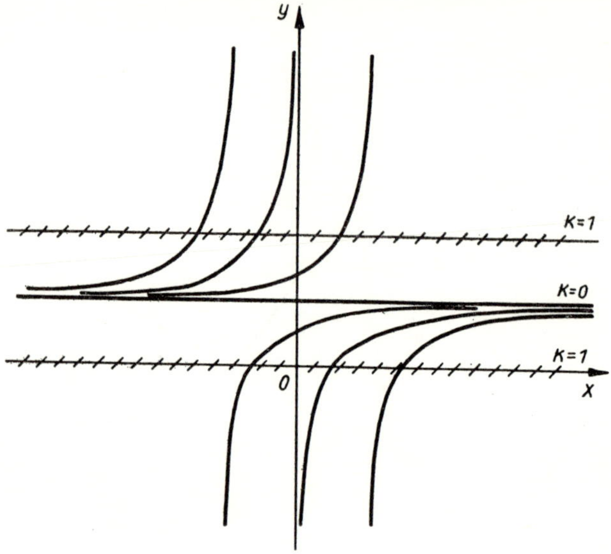

Answer to Problem 44

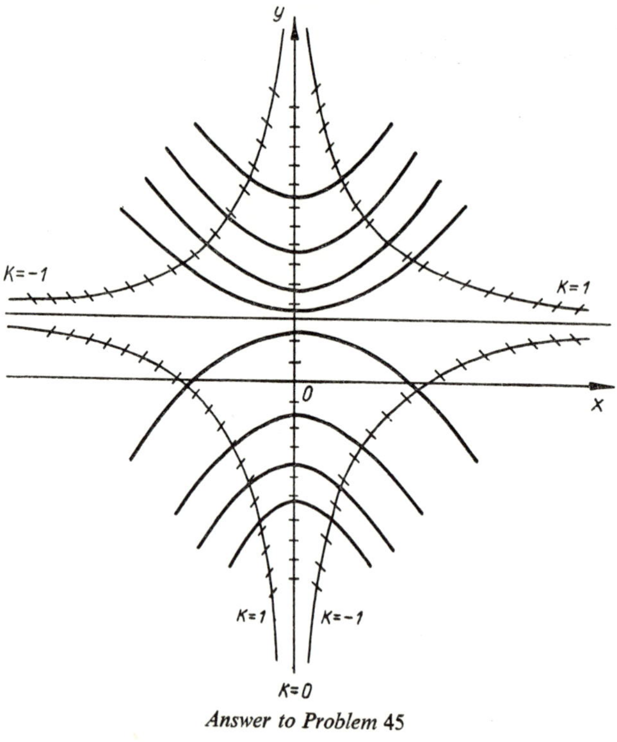

Answer to Problem 45

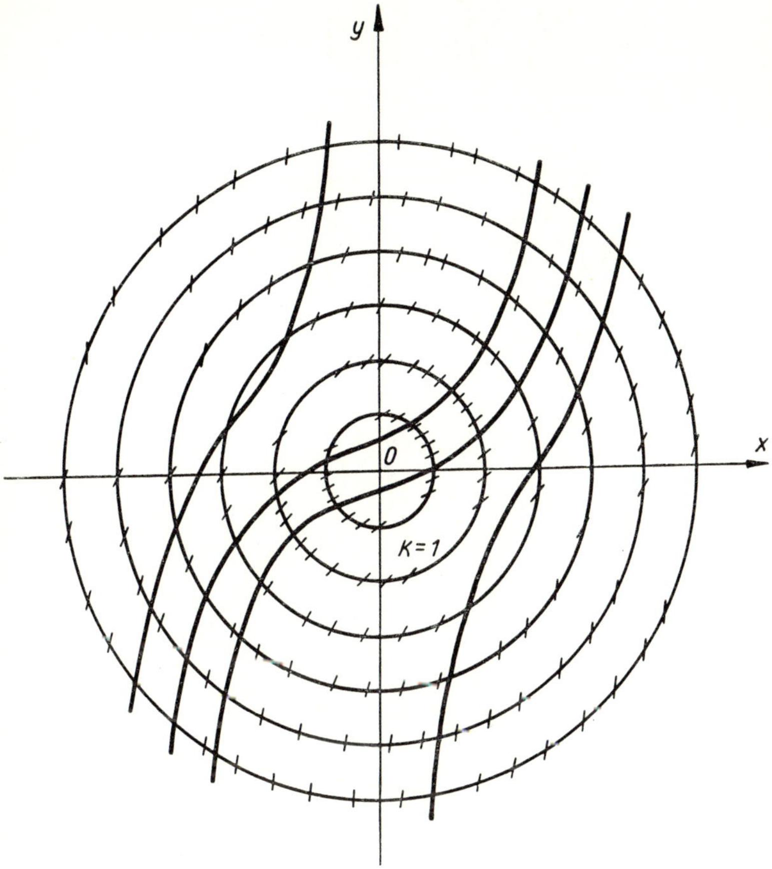

Answer to Problem 46

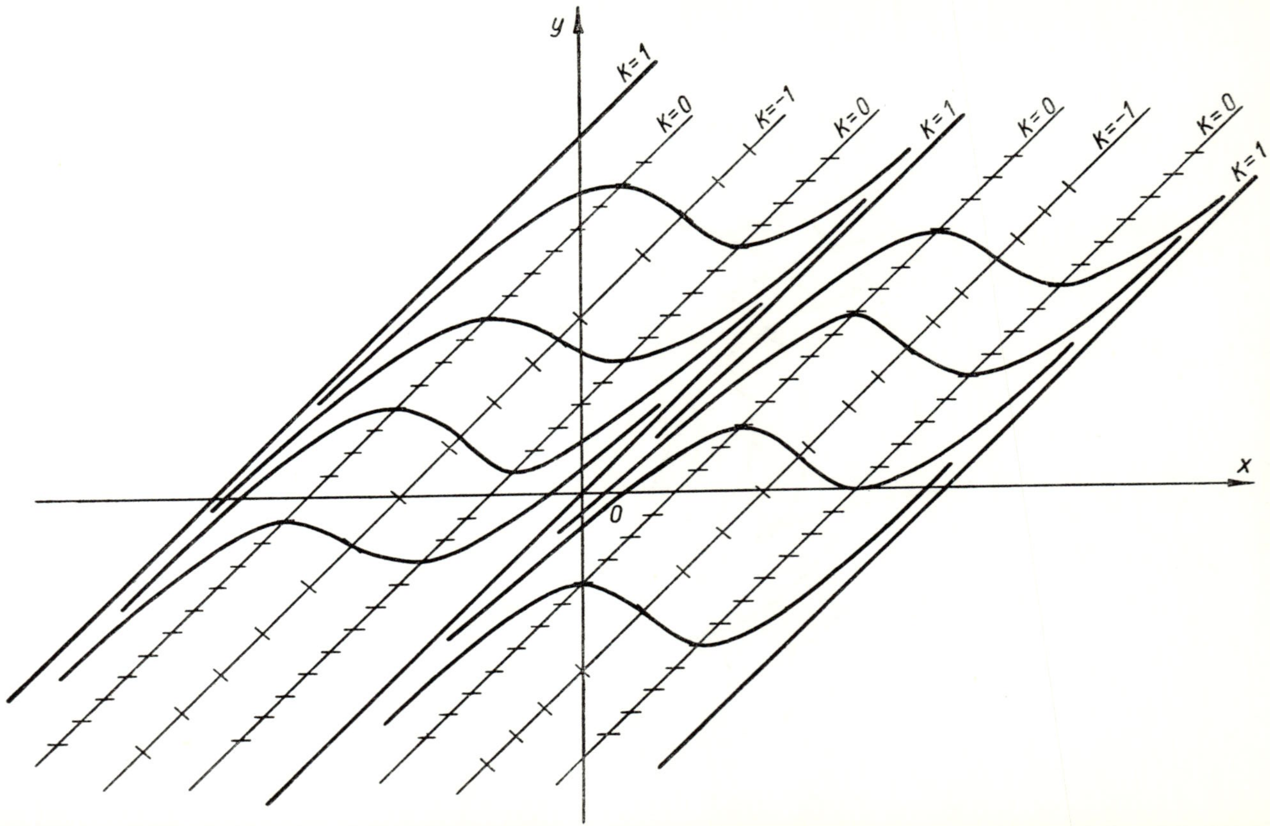
y
x
0
K=1
K=0
K=-1
K=0
K=1
K=0
K=-1
K=0
K=1

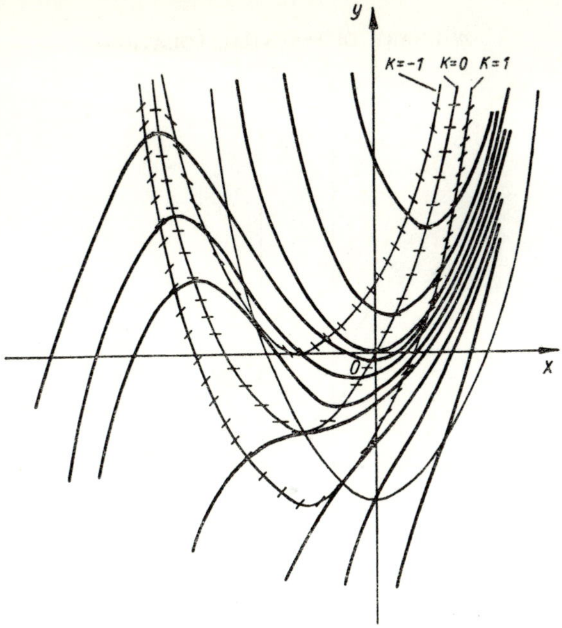

Answer to Problem 48

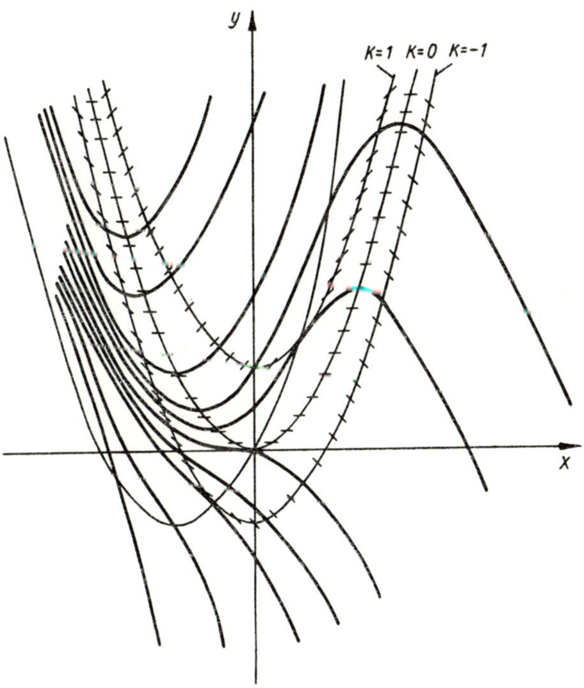

Answer to Problem 49

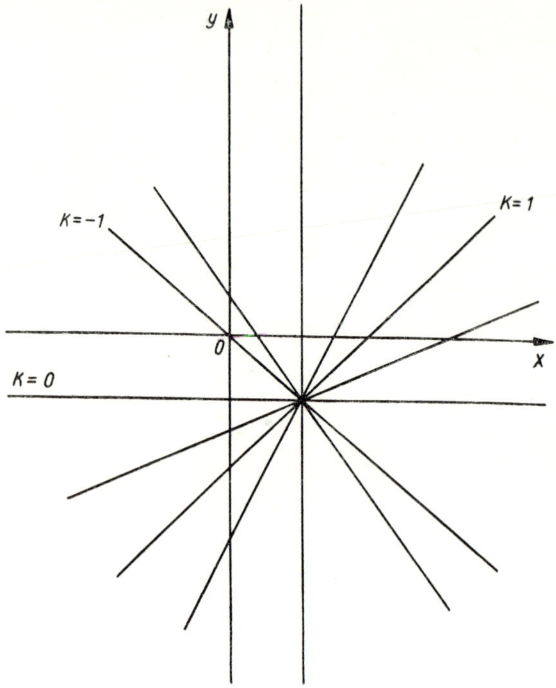

Answer to Problem 50

60. 2·337.

61. 20·77.

67. $x + y = C(1 - xy)$.

68. $x^2(1 + y^2) = C$.

69. $(x + y)(x - y - 2) + 2 \ln\left|\frac{1 + x}{1 - y}\right| = C$.

70. $y = \tan \ln Cx$.

71. $\sqrt{1 + x^2} + \sqrt{1 + y^2} = C$.

72. $\sqrt{1 - x^2} + \sqrt{1 - y^2} = 1; \quad y = 1$.

73. $e^x = C(1 - e^{-y})$.

74. $y = 1$.

75. $a^x + a^{-y} = C$.

76. $1 + e^y = C(1 + x^2)$.

77. $(1+y)e^{-y} = \ln\dfrac{1+e^x}{2} + 1 - x.$

78. $\frac{1}{2}e^{2x} - e^y - \ln\sqrt{1+y^2} - \arc\tan y = C.$

79. $(x^2 - 2x + 2)(y^2 + 1)e^{2\arc\tan y} = C.$

80. $x + C = \cot\left(\dfrac{y-x}{2} + \dfrac{\pi}{4}\right).$

81. $b(ax + by + c) + a = Ce^{bx}.$

82. $x + y = a\tan\left(C + \dfrac{y}{a}\right).$

83. $xe^{\frac{y^2}{x}} = C.$

84. $\ln\dfrac{\sqrt{1+y^2}}{y+\sqrt{1+y^2}} = \dfrac{x}{\sqrt{1+x^2}} + C.$

85. $2x^3y^3 = 3a^2x^2 + C.$

86. $\dfrac{1}{1-xy} + \dfrac{1}{2}\ln x = C.$

87. $Cy^2 = e^{xy - \frac{1}{xy}}.$

88. $3x^2 - 12x + 2x^3y^3 + 6xy = C.$

89. $\dfrac{x^3}{3} - x^2 + 2x + \dfrac{y^3}{3x^3} - \dfrac{4y}{x} = C.$

90. $x = \dfrac{(x+y)^{n-m+1}}{n-m+1} + \dfrac{(x+y)^{p-m+1}}{p-m+1} + C.$

$(n - m \neq -1, \quad p - m \neq -1).$

91. $y^3 = Cx - \ln x - 1.$

92. $2x^2 + (2x\ln y + 1)^2 = C; \quad x = 0.$

93. $y = a + \dfrac{Cx}{ax+1}.$

94. $y = a\tan\sqrt{\dfrac{a}{x} - 1}.$

95. $\ln\left|\tan\dfrac{y}{4}\right| = C - 2\sin\dfrac{x}{2},$

$y = 2\pi k \ (k = 0, \pm 1, \pm 2, \ldots).$

96. $y' = 3y; \ y = -2e^{3x}.$

97. $\int_0^x y\,dt = a^2 \ln \frac{y}{a}$, $y = \frac{a^2}{a - x}$ (hyperbola).

98. $\frac{dv}{dt} = 20\,\frac{t}{v}$; $v = \sqrt{725}\,.\,10$ cm/sec.

100. $m\frac{dv}{dt} = kv^2$; $t = \frac{h(v_1 - v_0)}{v_0 v_1 \ln \frac{v_1}{v_0}} = \frac{3}{40 \ln 2{\cdot}5}$ sec.

101. $m\frac{dv}{dt} = -kv$; $t = -\frac{5 \ln 10}{\ln 0{\cdot}8}$ sec.

103. $\frac{dT}{dt} = k(T - T_0)$; $T = 20 + 80(\frac{1}{2})^{\frac{t}{20}}$; 60 min.

104. $y' = n\frac{y}{x}$; $y = Cx^n$.

105. $\frac{ds}{dt} = ks$; $s = 25.2^{\frac{t}{5}}$.

106. 18·1 kg; $\frac{dx}{dt} = k\left(\frac{1}{3} - \frac{x}{300}\right)$,

where k is the coefficient of proportionality.

107. 5·2 kg; $\frac{dx}{dt} = kx\left(\frac{10 - x}{90} - \frac{1}{3}\right)$.

108. $\frac{1}{1{,}024}$; $\frac{dx}{dl} = -kx$.

109. 0·82 kg; $\frac{ds}{dt} = ks(s + 6)$.

110. 37·3 min.

111. $T = \frac{2}{3}x$; 864,000 cal.

114. $y = 0$, unique solution for $\alpha \leqslant 1$.

116. $y = \frac{\pi}{2}(2n + 1)x + C$ $(n = 0, \pm 1, \pm 2, \ldots)$.

117. $y = C$.

118. $y = (-1)^n(x \arcsin x + \sqrt{1 - x^2}) + \pi n x + C$, $(n = 0, \pm 1, \pm 2, \ldots)$.

119. $y = e^x + C$.

120. $y = \pi n x + C$, $(n = 0, \pm 1, \pm 2, \ldots)$.

121. $y = x(\ln x - 1) + C$.

122. $y = x \arctan x - \frac{1}{2}\ln(1 + x^2) + \pi n x + C.$

123. $y = \arcsin\left(\frac{\sqrt{3}}{2} + \frac{1}{x}\right) + 5\pi.$

124. $y = \operatorname{arc\,cot}\left(\frac{2}{x} + \frac{1}{\sqrt{3}}\right) + 3\pi.$

125. $y = 2\arctan\left(1 - \frac{1}{2x^2}\right) + \frac{9}{2}\pi.$

126. $y = \frac{1}{2}\arctan\left(\frac{\pi}{2} + \arctan x\right) + \frac{7}{2}\pi.$

127. $y = 0.$

128. $y = 1.$

129. $y = -\pi.$

130. $y = \operatorname{arc\,cot}\frac{2}{x} + \frac{9}{4}\pi.$

131. $y^2 - 3xy + 2x^2 = C.$

132. $2Cy = C^2x^2 + 1; \quad y = \pm x.$

133. $(x + y)(x^3 + y^3) = C.$

134. $(y - x)^8(y - 2x)^9 = C(y + 2x)^5.$

135. $C(y^2 - x^2) = y^3.$

136. $y^2 = Cxe^{\frac{x^2}{y^2}}.$

137. $y + \sqrt{y^2 - x^2} = Cx^3 e^{\frac{y(y+\sqrt{y^2-x^2})}{x^2}}.$

138. $fy^3 + 3cxy^2 + 3bx^2y + ax^3 = C.$

139. $x^2 = y^4 + Cy^6.$

140. $y^2 = x \ln Cy^2.$

141. $C^2x^2 = 1 + 2Cy; \quad C^2 - x^2 = 2Cy.$

142. $(x - 1)(3x + 2y - 1) = C.$

143. $x + y + 1 = Ce^{\frac{2x+y}{3}}.$

144. $(x + y - 1)^5(x - y - 1)^2 = C.$

145. $\sqrt{x^2y^4 + 1} = Cy^2 + 1.$

146. $y(x^2y - 1)^2 = C.$

147. $\arctan\frac{y^3}{x} = \frac{1}{2}\ln(x^2 + y^6) + \ln C.$

148. $x^2(x^2y + \sqrt{1 + x^4y^2}) = C.$

149. $\sin x + 2y \ln |y| - Cy = 0.$

150. $x = Ce^{-\sin\frac{y}{x}}.$

151. $C\sqrt{x^2 + y^2} = y^2 + 2x^2.$

152. $ye^{\sqrt{\frac{x}{y}}} = C; \quad x = 0.$

153. $x^2 + y^2 = Cx.$

154. $y = \frac{1}{2}\left(Cx^{1-k} - \frac{1}{C}x^{1+k}\right).$

155. Parabola $y^2 = 2Cx + C^2.$

156. $y = \frac{1}{2}\left(Cx^2 - \frac{1}{C}\right).$

157. $x^2 + y^2 = Cx^4.$

158. $y = Ce^{-2x} + \frac{1}{4}(2x^2 + 2x - 1).$

159. $y = C\sqrt{x^2 + 2x - 1} + x.$

160. $y = C \ln x + x^3.$

161. $y = C\sqrt{|a^2 - x^2|} + x.$

162. $y = C\sqrt{x} + x^2.$

163. $y = C(x + 1)^2 + \frac{1}{2}(x + 1)^4.$

164. $x = 8 \sin^2 \frac{y}{2} + Ce^{-\cos y}.$

165. $y = (x^2 + C)e^{x^2}.$

166. $y = \frac{Cx}{x^3 + 1} + \frac{1}{x}.$

167. $y = 2e^{-\sin x} + \sin x - 1.$

168. $y = Cx \ln x + \sqrt{x}.$

169. $y^3 = Cx^2 + x^3.$

170. $y^4 = C\sqrt{x} + \sqrt{x + 1}.$

171. $\frac{1}{x} = 2 - y^2 + Ce^{-\frac{y^2}{2}}.$

172. $\frac{1}{y} = Ce^{x^2} + \frac{1}{x}.$

173. $x^2 + y^2 - a^2 = Cy.$

174. $\frac{1}{y^2} = C \sin x + x.$

175. $x^2 = Ce^y - y - 2.$

176. $\frac{1}{y} = C\sqrt{x^2 + x + 1} - \frac{x}{\sqrt{x^2 + x + 1}}.$

177. $y^3 = \frac{Cx}{x^2 - a^2} + x^2.$

178. $\frac{1}{y} = \frac{1}{\sqrt{1 + x^2}}\left(C - \frac{x}{2}\sqrt{1 + x^2} + \frac{1}{2}\ln|\sqrt{1 + x^2} + x|\right).$

179. $\frac{1}{y^2} = \frac{1}{2}(x + 1)^4 + C(x + 1)^2.$

180. $y^4 + 2x^2y^2 + 2y^2 = C.$

181. $x = y \ln y + \frac{C}{y}.$

182. $y = \frac{Cx}{x - 1} + x^2.$

183. $y = \frac{x}{\cos x}.$

184. $\sin y = x + Ce^{-x}.$

185. $\tan\frac{y}{2} = Ce^{-x} - x + 1.$

186. $y = (x + 1)^n(C + e^x).$

187. $\phi(x) = Cx^{-\frac{n-1}{n}}.$

188. $\tan y = \left(C + \frac{x^2}{2}\right) e^{-x^2}.$

189. $y = \sin x.$

190. $y = \cos \sqrt{x}.$

191. $y = 2^{\sin x}.$

192. $y = \frac{\sin x}{x}.$

193. $y = \frac{\sin x}{x}.$

194. $y = \arctan x.$

195. $y = e^{e^x} + \cos\frac{1}{x}.$

196. $y = x^{-x}.$

197. $x^4 + x^2y^2 + y^4 = C.$

198. $x^3 + 3x^2y^2 + y^4 = C.$

199. $\sqrt{x^2 + y^2} + \ln|xy| + \frac{x}{y} = C.$

200. $x^3 \tan y + y^4 + \frac{y^3}{x^2} = C.$

201. $x^3y + x^2 - y^2 = Cxy.$

202. $\frac{\sin^2 x}{y} + \frac{x^2 + y^2}{2} = C.$

203. $x^3 + y^3 - x^2 - xy + y^2 = C.$

204. $y\sqrt{1 + x^2} + x^2y - y \ln x = C.$

205. $\sqrt{x^2 + y^2} + \frac{y}{x} = C.$

206. $x \sin y - y \cos x + \ln|xy| = C.$

207. $\tan xy - \cos x - \cos y = C.$

208. $y = x.$

209. $\sin(nx + my) + \cos(mx + ny) = C.$

210. $\arcsin\sqrt{x^2 + y^2} + \arcsin\frac{x}{y} + e^{\frac{x}{y}} = C.$

211. $\sin\frac{y}{x} - \cos\frac{x}{y} + x - \frac{1}{y} = C.$

212. $(x^2 + y^2)^2 + 2a^2(y^2 - x^2) = C.$

213. $1 + y^2 - x^2 = Cx;\quad \mu = \frac{1}{(1 + y^2 - x^2)^2}.$

214. $xy^2 - 2x^2y - 2 = Cx.$

215. $x^2 + y^2 = Cy^2e^{2x};\quad \mu = \frac{1}{y(x^2 + y^2)}.$

216. $\frac{y - 1}{\sqrt{x^2 + y^2}} = C;\quad \mu = \frac{1}{(x^2 + y^2)^{3/2}}.$

217. $x - \frac{y}{x} = C;\quad \mu = \frac{1}{x^2}.$

218. $x \ln |x| - y^2 = Cx; \quad \mu = \frac{1}{x^2}.$

219. $5 \arctan x + 2xy = C; \quad x = 0; \quad \mu = \frac{1}{1 + x^2}.$

220. $y^3 + x^3 (\ln x - 1) = Cx^2; \quad \mu = \frac{1}{x^4}.$

221. $2e^x \sin y + 2e^x(x - 1) + e^x(\sin x - \cos x) = C; \quad \mu = e^x.$

222. $x^2 - \frac{7}{y} - 3xy = C; \quad \mu = \frac{1}{y^2}.$

223. $(x + y^2)^2C = x - y^2; \quad \mu = \frac{1}{(x + y^2)^3}.$

224. $y = \frac{x^2}{2} + C, \quad y = Ce^x - x - 1.$

225. $(y - C)^2 = 4Cx.$

226. $(y - C)^2 = x^3.$

227. $\ln Cy = x \pm 2e^{\frac{x}{2}}, \quad y = 0.$

228. $x^3y^2 - Cxy(x + 1) + C^2 = 0.$

229. $y = \frac{C}{2} x^2 + \frac{1}{2C}, \quad y = \pm x.$

230. $(x - C)^2 + (y - C)^2 = C^2; \quad x = 0; \quad y = 0.$

231. $4e^{-\frac{y}{3}} = (x + 2)^{\frac{4}{3}} + C.$

232. $y = \frac{x^2}{2} + C, \quad y = -\frac{x^2}{2} + C, \quad y = Ce^x.$

233. $\left.\begin{array}{l} x = e^p(p + 1) + C \\ y = p^2e^p \end{array}\right\}; \quad y = 0.$

234. $Cx = \ln Cy; \quad y = e^x.$

235. $\left.\begin{array}{l} x = \ln p + \sin p \\ y = C + p(1 + \sin p) + \cos p \end{array}\right\}.$

236. $\left.\begin{array}{l} x = p^2 - 2p + 2 \\ y + C = \frac{2}{3}p^3 - p^2 \end{array}\right\}.$

237. $y = (\sqrt{C + 2x} - 1)e^{\sqrt{C+2x}-1}.$

238. $\left.\begin{array}{l} x + C = 2 \arctan p - \ln \left(\frac{1 + \sqrt{1 - p^2}}{p}\right) \\ y = \arcsin p + \ln (1 + p^2) \end{array}\right\}; \; y = 0.$

239. $\left.\begin{array}{l} x = e^p + C \\ y = (p-1)e^p \end{array}\right\}; \quad y = -1.$

240. $\left.\begin{array}{l} x = \dfrac{e^{\frac{1}{p}}}{p^2} \\ y = C + e^{\frac{1}{p}}\left(1 + \dfrac{1}{p}\right) \end{array}\right\}.$

241. $y + C = \sqrt{x - x^2} + \arcsin \sqrt{x}.$

242. $\left.\begin{array}{l} x = a\cos^3 t \\ y + C = -a\sin^3 t \end{array}\right\}, \quad (p = \tan t).$

243. $\left.\begin{array}{l} x = 5(\frac{1}{3}\tan^3 t - \tan t + t) + C \\ y = a\sin^5 t \end{array}\right\}.$

244. $\left.\begin{array}{l} x = -\dfrac{2}{t} + \ln\left|\dfrac{t+1}{t-1}\right| - 2\arctan t \\ y = \dfrac{t^2}{1 - t^4} \end{array}\right\}; \quad y = 0.$

$(y' = ty)$

245. $\left.\begin{array}{l} x = p + \sin p \\ y + C = \frac{1}{2}p^2 + p\sin p + \cos p \end{array}\right\}.$

246. $\left.\begin{array}{l} x + C = \ln p + \sin p + p\cos p \\ y = p + p^2\cos p \end{array}\right\}.$

247. $\left.\begin{array}{l} x = 2Cp - \ln p - 2 \\ y = Cp^2 - p \end{array}\right\}.$

248. $\left.\begin{array}{l} x = \dfrac{C}{p^2} - \dfrac{1}{p} \\ y = \dfrac{2C}{p} + \ln p - 2 \end{array}\right\}.$

249. $\left.\begin{array}{l} x = 2(1-p) + Ce^{-p} \\ y = [2(1-p) + Ce^{-p}](1+p) + p^2 \end{array}\right\}.$

250. $\left.\begin{array}{l} x = \dfrac{C}{p^2} - \dfrac{\cos p}{p^2} - \dfrac{\sin p}{p} \\ y = \dfrac{2C}{p} - \dfrac{2\cos p}{p} - \sin p \end{array}\right\}.$

251. $$\left.\begin{aligned} x &= \frac{Cp^2 + 2p - 1}{2p^2(p-1)^2} \\ y &= \frac{Cp^2 + 2p - 1}{2(p-1)^2} - \frac{1}{p} \end{aligned}\right\}.$$

252. $$\left.\begin{aligned} x &= \frac{C}{p^3} - 2e^p\left(\frac{1}{p} - \frac{2}{p^2} + \frac{2}{p^3}\right) \\ y &= \frac{3C}{2p^2} - 2e^p\left(1 - \frac{3}{p} + \frac{3}{p^2}\right) \end{aligned}\right\}.$$

253. $y = Cx + \dfrac{a}{C^2};\quad 4y^3 = 27ax^2.$

254. $y = Cx + C^2;\quad y = -\dfrac{x^2}{4}.$

255. $y = Cx - \dfrac{C-1}{C};\quad (y+1)^2 = 4x.$

256. $y = Cx + a\sqrt{1 + C^2};\quad x^2 + y^2 = a^2.$

257. $y = Cx + \dfrac{aC}{\sqrt{1 + C^2}};\quad x^{\frac{2}{3}} + y^{\frac{2}{3}} = a^{\frac{2}{3}}.$

258. $x = Cy + C^2;\quad 4x = -y^2.$

259. $xy = \pm a^2.$

260. $x^{\frac{2}{3}} + y^{\frac{2}{3}} = a^{\frac{2}{3}}.$

261. $y + xy' = 0.$

262. $x^2 + y^2 - 2xyy' = 0.$

263. $xy' = y \ln y'.$

264. $y'^2 + y' - xy' + y = 0.$

265. $y'' - 2y' + y = 0.$

266. $yy'^2 + 2xy' - y = 0.$

267. $y''' = 0.$

268. $y''' + \dfrac{3}{x}y'' = 0.$

269. $y''^2 = (1 + y'^2)^3.$

270. $y'' - y = 0.$

271. $y'' + y = 0.$

272. $y^2 - 2bx = b^2,\quad (b \geqslant 0).$

273. $\frac{x^2}{b^2} - \frac{y^2}{c^2 - b^2} = 1, \quad (b < c).$

274. $2x + \sigma y^2 = C.$

275. $\sin y = be^{-x}.$

276. $y^2 = 2bx.$

277. $xy = C.$

278. $y = ax, \quad \text{if} \quad k = 2; \quad \frac{1}{x^{k-2}} - \frac{1}{y^{k-2}} = \frac{1}{b^{k-2}}, \quad k \neq 2.$

279. $x^2 + y^2 = 2bx.$

280. $xy^3 = b^4.$

281. $\rho = C(1 - \cos \phi).$

282. $y = Ce^{-\frac{x}{2}}.$

283. $x = ay^2 + Cy\sqrt{1 - y^2}.$

284. $y = x - \frac{1}{x + C}.$

285. $y = C\frac{\sin x}{x} + \cos x.$

286. $y^{1-n} = 2 \sin x + \frac{2}{n - 1} + Ce^{(n-1) \sin x}.$

287. $x^4 - 6x^2y^2 + y^4 = C.$

288. $(y - x)^5(y - 3x)^9 = C(y - 2x)^{12}.$

289. $6y + 12y^3 - 9x^2y^2 + 2x^3 = C.$

290. $2 + xy \ln^2 x = Cxy, \quad \mu = \frac{1}{x^2y^2}.$

291. $y = Ce^{-x^2} + (\sin x - x \cos x)e^{-x^2}.$

292. $(C - \ln x)(1 - xy) = 2.$

293. $x = Ce^{2y} + \frac{1}{2}y^2 + \frac{1}{2}y + \frac{1}{4}.$

294. $y + C = 2x - \frac{x^2}{2} + 2 \ln |1 - x|.$

295. $x^4 = Cy^2e^{\frac{1}{y}} + 2y^2 + 2y + 1.$

296. $y^2 = Cx^2 + x^4.$

297. $y(y - 2x)^3 = C(y - x)^2.$

298. $y = C(2x - 1) + \frac{1}{x}$.

299. $x + y - 1 = Ce^{\frac{2x+2}{x+y-1}}$.

300. $\ln\left|\tan\frac{y}{4}\right| = C - 2\cos\frac{x}{2}$.

301. $y = C(3x^2 - 2x) + \frac{2}{x}$.

302. $y^3 = Cx^3 + x^4$.

303. $$x + C = \frac{\frac{a}{b}(ax + by + c) + \ln\left[\frac{a}{b} + \tan(ax + by + c)\right]}{b\left(1 + \frac{a^2}{b^2}\right)} - \frac{\frac{1}{2}\ln\left[1 + \tan^2(ax + by + c)\right]}{b\left(1 + \frac{a^2}{b^2}\right)}.$$

304. $x + ye^{\frac{x}{y}} = 1 + e$.

305. $\ln|x| - \frac{y^2}{2x^2} = C$.

306. $(ay + bx)^2(x^2 + y^2 - C) + x^2y^2 = 0$.

307. $x^2y + 2x = Cy$.

308. $x^2 + y^2 = Ce^{-x}$.

309. $\frac{1}{2}\ln\frac{x^2 + x + 1}{y^2 - y + 1} - \sqrt{3}\left(\arctan\frac{2x + 1}{\sqrt{3}} + \arctan\frac{2y - 1}{\sqrt{3}}\right) = C$.

310. $x = y^2(1 + Ce^{\frac{1}{y}})$.

311. $\sin x + 2y\ln|y| - Cy = 0$.

312. $3e^{-2y} = Ce^{-2x} - 2e^{x}$.

313. $x^4 + y^2 = C(x^2 + y)$.

314. $\frac{1}{x} = \frac{C}{y^2} + \frac{y^n}{n + 2}, \quad n \neq -2$.

315. $$x\sqrt{1 + x^2} + y\sqrt{1 + y^2} + \ln|x + \sqrt{1 + x^2}| + \ln|y + \sqrt{1 + y^2}| + 2nxy = n\sqrt{1 + n^2} + \ln|n + \sqrt{1 + n^2}|.$$

316. $\frac{3}{7}y - \frac{4}{7}x + \frac{1}{49} + \frac{1}{49}(4a^2 - 3b^2)\ln|7(x + y) + a^2 + b^2| = C$.

317. $y^2 - Cx^2 = -\dfrac{bC}{1 + aC}$.

345. $y = (x - 3)e^x + \dfrac{x^2}{2} + 2x + 3$.

346. $y = \dfrac{x^5}{120} + C_1x^3 + C_2x^2 + C_3x + C_4$.

347. $y = \dfrac{x^4}{24}\ln x - \dfrac{13}{288}x^4 + \dfrac{x^2}{8} - \dfrac{x}{9} + \dfrac{1}{32}$.

348. $y = \dfrac{x^4}{24} - \sin x + C_1x^2 + C_2x + C_3$.

349. $y = -\dfrac{2x + 3}{12(x + 2)^2} + \dfrac{x^2 - 6x + 20}{324}$.

350. $y + C_2 = \dfrac{6}{5}(x + C_1) + \dfrac{5}{12}(x + C_1)^3$.

351. $y = (1 + C_1^2)\ln|x + C_1| - C_1x + C_2$.

352. $\left.\begin{aligned} x + C_1 &= \dfrac{1}{2}\ln t + \dfrac{3}{4t^2} \\ y + C_2 &= \dfrac{1}{4}t + \dfrac{3}{4t^3} \end{aligned}\right\}, \quad t = y''$.

353. $y = (C_1x - C_1^2)e^{\frac{x}{C_1}+1} + C_2$.

354. $y + C_2 = \ln\left|\tan\left(\dfrac{x}{2} + C_1\right)\right|$.

355. $y = C_3 + C_2x - \sin(x + C_1)$.

356. $\left.\begin{aligned} x + C_2 &= z(2\ln z - 1) \\ y + C_1 &= z^2\ln z \end{aligned}\right\}, \quad z = y'$.

357. $\left.\begin{aligned} x &= z^2 + 1 \\ y &= \dfrac{4}{15}z^5 + C_1z^2 + C_2 \end{aligned}\right\}, \quad z = y''$.

358. $y = C_1x(x - C_1) + C_2; \quad y = \dfrac{x^3}{3} + C$.

359. $y = C_2\left(xe^{C_1x} - \dfrac{1}{C_1}e^{C_1x}\right) + C_3$.

360. $\left.\begin{aligned} x + C_2 &= e^z(z + 1) \\ y + C_1 &= z^2e^z \end{aligned}\right\}, \quad z = y'$.

361. $y = \frac{2}{5} x^2\sqrt{2x} - \frac{16}{5}$.

362. $y = \frac{1}{12}(x^3 + 6x^2) + C_1 x \ln |x| + C_2 x + C_3$.

363. $x = C_1 y^2 + C_2 y + C_3$.

364. $y = C_2 + \frac{3}{7}(x + C_1)^{\frac{7}{3}} - \frac{3}{4} C_1(x + C_1)^{\frac{4}{3}}$.

365. $y = C_3 + C_2 x + C_1 \ln x - \frac{1}{2x}$.

366. $y = C_2 - \frac{1}{2}\sqrt{1 - C_1^2 x^2} + \frac{1}{2} C_1 x\sqrt{1 - x^2} + \frac{1}{2} C_1 \arcsin x$.

367. $y = \frac{1}{2} x \ln |x| + C_1 \ln |x - 1| + C_2 x + C_3$.

368. $C_1 y^2 - 1 = (C_1 x + C_2)^2$.

369. $y = \frac{C_1}{2}\left(e^{\frac{x+C_2}{C_1}} + e^{-\frac{x+C_2}{C_1}}\right)$.

370. $C_1\sqrt{C_1} x + C_2 = \sqrt{C_1 y(C_1 y - 1)} + \ln(\sqrt{C_1 y} + \sqrt{C_1 y - 1})$

371. $x + C_2 = \frac{1}{C_1} \ln \frac{\sqrt{C_1^2 + ae^y} - C_1}{\sqrt{C_1^2 + ae^y} + C_1}$.

372. $3x + C_2 = 4\sqrt{C_1 + \sqrt{y}}(\sqrt{y} - 2C_1)$.

373. $C_2^2(x - C_1) = \left(C_2 y^{\frac{2}{3}} + 2\right)\sqrt{C_2 y^{\frac{2}{3}} - 1}$.

374. $(x - C_1)^2 = 4C_2(y - C_2)$.

375. $y = \sqrt{2x - x^2}$.

376. $2C_2 y^2 = 2C_1 C_2 + C_2^2 e^{2x} + (C_1^2 - 1)e^{-2x}$.

377. $C_1 x + C_2 = \ln\left|\frac{y}{y + C_1}\right|$.

378. $y = C_2 e^{C_1 x}$.

379. $y = -\ln |x - 1|$.

380. $y \cos^2(x + C_1) = C_2$.

381. $(x - C_1)^2 + (y - C_2)^2 = 1$.

382. $y = C_2 e^{\frac{1}{3}(x^2 + C_1)^{\frac{3}{2}}}$.

383. $y = x$.

384. $y = C_3 e^{x(\frac{1}{2}\ln^2 x + C_1 \ln x + C_2)}$.

385. $\dfrac{d^2x}{dt^2} = -\dfrac{k}{x^2}$, ($x$ is the distance of the body from the centre of the earth); $t \approx 122$ hours.

386. $m\dfrac{d^2x}{dt^2} = \dfrac{k}{x^3}$; $x^2 = \dfrac{a^2}{C_1}(t + C_2)^2 + C_1$; $a^2 = \dfrac{k}{m}$.

387. $m\dfrac{d^2x}{dt^2} = mg - k\left(\dfrac{dx}{dt}\right)^2$; $x = \dfrac{m}{k}\ln\dfrac{e^{\alpha t} + e^{-\alpha t}}{2}$, $\alpha = \sqrt{\dfrac{kg}{m}}$.

388. $Cx = y^{2k-1}$ $(k > \frac{1}{2})$.

389. $(x + C_1)^2 + (y + C_2)^2 = R$, where $R = \text{const.}$

390. Yes. **391.** No. **392.** No. **393.** Yes.

394. Yes. **395.** Yes. **396.** No. **397.** No.

398. No. **399.** No. **400.** No. **401.** No.

402. No. **403.** No. **404.** Yes. **405.** No.

406. 1.

407. $-\dfrac{2}{x}$ $(x \neq 0)$.

408. 0.

409. e^{-2x}.

410. 0.

411. $-8\sin^3 x$.

412. $-\dfrac{1}{\sqrt{2}}$.

413. $\dfrac{\pi}{2\sqrt{\pi^2 - x^2}}$ $(|x| < \pi)$.

414. 0.

415. 0.

416. $1 - \ln x$ $(x > 0)$.

417. $\dfrac{x-1}{x^3} e^{\frac{1}{x}}$ $(x \neq 0)$.

418. $-e^{2x}$.

419. $-2e^{-6x}$.

420. 1.

421. 1.

426. $y'' + 3y' + 2y = 0.$

427. $2y'' - 3y' - 5y = 0.$

428. $y''' + 3y'' + 2y' = 0.$

429. $y^{\text{IV}} + 2y'' + y = 0.$

430. $y''' = 0.$

431. $y'' - 3y' + 2y = 0, \quad y = C_1e^x + C_2e^{2x}.$

432. $y'' - 2y' + y = 0, \quad y = C_1e^x + C_2xe^x.$

433. $y'' - 6y' + 13y = 0, \quad y = e^{3x}(C_1 \cos 2x + C_2 \sin 2x).$

434. $y''' - 3y'' + 3y' - y = 0, \quad y = (C_1 + C_2x + C_3x^2)e^x.$

435. $y'' - y = 0.$

436. $y'' - y' = 0.$

437. $y'' - 4y' + 4y = 0.$

438. $y'' + 9y = 0.$

439. $y'' = 0.$

440. $y''' - 6y'' + 11y' - 6y = 0.$

441. $y''' - 3y'' + 3y' - y = 0.$

442. $y''' - 4y'' + 5y' - 2y = 0.$

443. $y''' - y'' = 0.$

444. $y''' + y' = 0.$

445. $y''' - 2y'' + y' - 2y = 0.$

446. $y''' + y'' + 2y' = 0.$

447. $y = C_1e^x + C_2e^{-x}.$

448. $y = C_1e^{2x} + C_2e^{-\frac{4}{3}x}.$

449. $y = e^x(1 + x).$

450. $y = e^{-x}(C_1x + C_2).$

451. $y = 4e^x + 2e^{3x}.$

452. $y = C_1e^{-x} + C_2e^{-2x} + C_3e^{-3x}.$

453. $y = C_1e^{(1-\sqrt{3})x} + C_2e^{(1+\sqrt{3})x}.$

454. $y = C_1 + C_2x + C_3x^2 + C_4x^3 + e^{-x}(C_5 + C_6x).$

455. $y = e^x\left(C_1 \cos \frac{x}{2} + C_2 \sin \frac{x}{2}\right).$

456. $y = C_1e^{2x} + e^{-x}(C_2 \cos \sqrt{3}x + C_3 \sin \sqrt{3}x)$.

457. $y = (C_1 + C_2x)e^{-x} + (C_3 \cos 2x + C_4 \sin 2x)e^{-x}$.

458. $y = e^x \sin x$.

459. $y = e^x(\cos \sqrt{2}x + \sqrt{2} \sin \sqrt{2}x)$.

460. $y = C_1e^x + C_2e^{-x} + e^{-x}(C_3 \cos 2x + C_4 \sin 2x)$.

461. $y = C_1e^x + C_2e^{-2x} + (C_3 + C_4 \cos x + C_5 \sin x)e^{-x}$.

462. $y = C_1e^x + C_2e^{-x} + C_3e^{-2x}$.

463. $y = C_1 + C_2e^x \cos x + C_3e^x \sin x$.

464. $y = C_1e^{-x} + C_2e^x + C_3 \cos x + C_4 \sin x$.

465. $y = C_1 + C_2x + C_3x^2 + \ldots + C_{10}x^9$.

466. $y = C_1e^{-x} + C_2xe^{-x} + C_3e^{2x}$.

467. $y = C_1 + C_2e^x + C_3e^{\frac{x}{2}}$.

468. $y_{p.n.} = A_1x^2 + A_2x + A_3$.

469. $y_{p.n.} = A_1x^3 + A_2x^2 + A_3x$.

470. $y_{p.n.} = A_1x^4 + A_2x^3 + A_3x^2$.

471. $y_{p.n.} = (A_1x + A_2)e^{-x}$.

472. $y_{p.n.} = (A_1x^2 + A_2x)e^{-x}$.

473. $y_{p.n.} = (A_1x^3 + A_2x^2)e^{-x}$.

474. $y_{p.n.} = A \sin x + B \cos x$.

475. $y_{p.n.} = x(A \sin x + B \cos x)$.

476. $y_{p.n.} = x(A \sin 2x + B \cos 2x)$.

477. $y_{p.n.} = x(A \sin kx + B \cos kx)$.

478. $y_{p.n.} = (A \sin x + B \cos x)e^{-x}$.

479. $y_{p.n.} = x(A \sin x + B \cos x)e^{-x}$.

480. $y_{p.n.} = A_1x^2 + A_2x + A_3$.

481. $y_{p.n.} = A_1x^3 + A_2x^2 + A_3x$.

482. $y_{p.n.} = A_1x^4 + A_2x^3 + A_3x^2$.

483. $y_{p.n.} = A_1x^5 + A_2x^4 + A_3x^3$.

484. $y_{p.n.} = (A_1 \sin x + B_1 \cos x)x$.

485. (*a*) $y_{p.n.} = x(A_1e^{-x} + A_2e^x)$.

(*b*) $y_{p.n.} = A_1e^{-x} + A_2e^x$.

(c) $y_{p.n.} = A_1x^2e^{-x} + A_2xe^x$.
(d) $y_{p.n.} = A_1x^3e^{-x} + A_2e^x$.
(e) $y_{p.n.} = A_1xe^{-x} + A_2x^2e^x$.
(f) $y_{p.n.} = A_1e^{-x} + A_2x^3e^x$.

486. (a) $y_{p.n.} = (A_1x^2 + A_2x + A_3)e^{kx}$.
(b) $y_{p.n.} = (A_1x^3 + A_2x^2 + A_3x)e^{kx}$.
(c) $y_{p.n.} = (A_1x^4 + A_2x^3 + A_3x^2)e^{kx}$.
(d) $y_{p.n.} = (A_1x^2 + A_2x + A_3)e^{kx}$.
(e) $y_{p.n.} = (A_1x^4 + A_2x^3 + A_3x^2)e^{kx}$.
(f) $y_{p.n.} = (A_1x^5 + A_2x^4 + A_3x^3)e^{kx}$.

487. (a) $y_{p.n.} = A\sin x + B\cos x$.
(b) $y_{p.n.} = x(A\sin x + B\cos x)$.

488. (a) $y_{p.n.} = x(A\sin 2x + B\cos 2x)e^{3x}$.
(b) $y_{p.n.} = x^2(A\sin 2x + B\cos 2x)e^{3x}$.

489. $y_{p.n.} = Ax$.

490. $y_{p.n.} = A_1x^3 + A_2x^2 + A_3x$.

491. $y_{p.n.} = Ae^x$.

492. $y_{p.n.} = Axe^{-7x}$.

493. $y_{p.n.} = (A_1x^3 + A_2x^2)e^{4x}$.

494. $y_{p.n.} = Ax^2e^{5x}$.

495. $y_{p.n.} = (A_1x^2 + A_2x)e^{\frac{3}{4}x}$.

496. $y_{p.n.} = A_1e^x + A_2e^{-2x}$.

497. $y_{p.n.} = (A_1x^2 + A_2x)e^{4x}$.

498. $y_{p.n.} = x(A\cos 5x + B\sin 5x)$.

499. $y_{p.n.} = x(A\cos x + B\sin x)$.

500. $y_{p.n.} = x(A\cos 4x + B\sin 4x)$.

501. $y_{p.n.} = (A\cos 2x + B\sin 2x)e^{8x}$.

502. $y_{p.n.} = x(A\cos 2x + B\sin 2x)e^{2x}$.

503. $y_{p.n.} = x(A\cos 2x + B\sin 2x)e^{-3x}$.

504. $y_{p.n.} = x(A\sin kx + B\cos kx)$.

505. $y_{p.n.} = A \qquad (A = \text{const})$.

506. $y_{p.n.} = A_1\cos x + B_1\sin x + A_2\cos 3x + B_2\sin 3x$.

507. $y_{p.n.} = A_1x + A_2\cos 8x + B_2\sin 8x$.

508. $y_{p.n.} = A_1x + A_2$.

509. $y_{p.n.} = A \qquad (A = \text{const})$.

510. $y_{p.n.} = Ax$.

511. $y_{p.n.} = Ax^2$.

512. $y_{p.n.} = A \qquad (A = \text{const})$.

513. $y_{p.n.} = Ax$.

514. $y_{p.n.} = Ax^2$.

515. $y_{p.n.} = Ax^3$.

516. $y_{p.n.} = Ax^2$.

517. $y_{p.n.} = Ae^{4x}$.

518. $y_{p.n.} = Ax^2e^{-x}$.

519. $y_{p.n.} = (A_1x^3 + A_2x^2)e^{-x}$.

520. $y_{p.n.} = A \sin 2x + B \cos 2x$.

521. $y_{p.n.} = A \sin x + B \cos x$.

522. $y_{p.n.} = (A_1x + A_2) \sin 2x + (B_1x + B_2) \cos 2x$.

523. $y_{p.n.} = x^2(A \sin nx + B \cos nx)$.

524. $y_{p.n.} = A \sin nx + B \cos nx$.

525. $y_{p.n.} = A \sin x + B \cos x$.

526. $y_{p.n.} = Ax^4e^x$.

527. $y_{p.n.} = (Ax^5 + Bx^4)e^x$.

528. $y = (C_1 + C_2x)e^{-x} - 2$.

529. $y = C_1 + C_2e^{-2x} - x$.

530. $y = C_1 \sin 3x + C_2 \cos 3x + 1$.

531. $y = C_1 + C_2x + C_3e^{-x} + \frac{1}{2}x^2$.

532. $y = C_1 + C_2x + C_3e^{\frac{7}{5}x} - \frac{3}{14}x^2$.

533. $y = C_1 + C_2x + C_3x^2 + C_4e^{6x} + \frac{1}{6}x^3$.

534. $y = C_1 + C_2x + C_3x^2 + C_4e^{-\frac{x}{3}} + \frac{x^3}{3}$.

535. $y = C_1 \cos x + C_2 \sin x + (C_3 + C_4x)e^x + 1$.

536. $y = (C_1 + C_2x)e^{2x} + \frac{x^2}{4} + \frac{x}{2} + \frac{3}{8}$.

537. $y = C_1 + C_2e^{-8x} + \frac{x^2}{2} - \frac{x}{8}.$

538. $y = (C_1 + C_2x)e^{kx} + \frac{e^x}{(k-1)^2}.$

539. $y = (C_1 + C_2x)e^{-2x} + 4x^2e^{-2x}.$

540. $y = C_1e^{-3x} + C_2e^{-x} - \frac{9}{2}xe^{-3x}.$

541. $y = C_1 + C_2e^{\frac{x}{7}} - 7x^2 - 98x.$

542. $y = C_1 + C_2e^{-3x} - \left(\frac{x^2}{2} + \frac{x}{3}\right)e^{-3x}.$

543. $y = C_1e^{-3x} + C_2e^{-2x} + (20x - 5x^2)\,e^{-2x}.$

544. $y = e^{-x}(C_1 \sin x + C_2 \cos x) + \frac{1}{2}x.$

545. $y = e^{-\frac{x}{2}}\left(C_1 \sin \frac{\sqrt{3}}{2}x + C_2 \cos \frac{\sqrt{3}}{2}x\right) + \left(\frac{x^2}{3} - \frac{x}{3} + \frac{1}{9}\right)e^x.$

546. $y = C_1e^{-(\sqrt{6}+2)x} + C_2e^{(\sqrt{6}-2)x} - \frac{12 \sin 2x + 16 \cos 2x}{25}.$

547. $y = C_1 \cos x + C_2 \sin x + x^2 \sin x + x \cos x.$

548. $y = (C_1 + C_2x)e^{mx} + \frac{(m^2 - n^2)\sin nx + 2mn \cos nx}{(m^2 + n^2)^2}$

549. $y = e^{-x}(C_1 \cos 2x + C_2 \sin 2x) - \frac{1}{4}xe^{-x} \cos 2x.$

550. $y = C_1 \cos ax + C_2 \sin ax + \frac{2 \cos mx + 3 \sin mx}{a^2 - m^2} \quad (m \neq a).$

551. $y = C_1 + C_2e^x - \frac{1}{2}e^x(\sin x + \cos x).$

552. $y = C_1 + C_2e^{-2x} + \frac{1}{5}e^{x}(6 \sin x - 2 \cos x)$

553. $y = e^{-2x}(C_1 \cos x + C_2 \sin x) + 5xe^{-2x} \sin x.$

554. $y = e^{-x}(C_1 \cos 2x + C_2 \sin 2x) - \frac{1}{4}xe^{-x} \cos 2x + \frac{x}{2}e^{-x}.$

555. $y = C_1 + C_2e^{-2x} - \left(\frac{x}{20} - \frac{7}{50}\right)\sin x - \left(\frac{x}{10} + \frac{1}{50}\right)\cos x.$

556. $y = C_1e^{2x} + \left(C_2 - x - \frac{x^2}{2}\right)e^x.$

557. $y = C_1e^x + C_2e^{-2x} + \frac{e^{4x}}{18}\left(x^2 - x + \frac{7}{18}\right).$

558. $y = C_1e^x + C_2e^{2x} + \frac{e^{3x}}{2}(x^2 - 2x + 2).$

559. $y = C_1e^x + C_2 \cos x + C_3 \sin x - x^2 - 3x - 1.$

560. $y = C_1e^x + C_2xe^x + C_3 \cos x + C_4 \sin x + \frac{x^2}{4} e^x.$

561. $y = (C_1 + C_2x)e^x + x^3 + 6x^2 + 18x + 24.$

562. $y = e^{\frac{3}{5}x}\left(C_1 \cos \frac{4}{5} x + C_2 \sin \frac{4}{5} x\right) + \frac{e^{2x}}{2} + 1{\cdot}3.$

563. $y = C_1 + C_2x + C_3 \cos x + C_4 \sin x + \frac{x^4}{12} + \frac{x^3}{6} - x^2.$

564. $y = \frac{x^4}{24} + C_1x^3 + C_2x^2 + C_3x + C_4 + \left(\frac{x^2}{2} - 4x + C_5\right) e^x.$

565. $y = \left(C_1 + \frac{x}{4} - \frac{x^3}{6}\right) \cos x + \left(C_2 + \frac{x^2}{4}\right) \sin x.$

566. $y = C_1e^{-x} + C_2xe^{-x} + e^{-x}(-x^2 \cos x + 4x \sin x + 6 \cos x).$

567. $y = C_1 + C_2e^{2x} + C_3e^{-2x} + \frac{1}{5} \cos x - \frac{x^3}{12} - \frac{x}{8} + \frac{e^{2x}}{32}(2x^2 - 3x).$

568. $y = C_1e^x + e^{-\frac{x}{2}}\left(C_2 \cos \frac{\sqrt{3}}{2} x + C_3 \sin \frac{\sqrt{3}}{2} x\right) + \frac{1}{2}(\cos x - \sin x).$

569. $y = \frac{x}{2} e^{-x} \sin x + xe^{-x} + e^{-x}(C_1 \cos x + C_2 \sin x).$

570. $y = \frac{1}{4} \cos x + (C_1 + C_2x)e^x + (C_3 + C_4x)e^{-x}.$

571. $y = \frac{1}{2}x \sin x + \frac{1}{8} \cos 3x + C_1 \cos x + C_2 \sin x.$

572. $y = C_1 \cos 2x + C_2 \sin 2x + \frac{1}{8}x - \frac{1}{32} x \cos 2x - \frac{1}{16} x^2 \sin 2x.$

573. $y = \left(\frac{x}{8} - \frac{1}{4}\right) e^x + (C_1 + C_2x)e^{-x} + \left(C_3 - \frac{x}{8}\right) \cos x + C_4 \sin x.$

574. $y = C_1 + C_2e^{-x} + \left(\frac{x^3}{3} - x^2 + \frac{5}{2}x\right) + \frac{1}{2} e^x + \frac{1}{20} \sin 2x - \frac{1}{10} \cos 2x.$

575. $y = C_1 + C_2x + C_3x^2 + C_4 \cos 2x + C_5 \sin 2x + \frac{1}{5} e^x + \frac{1}{24} x^3$
$+ \frac{3}{32} x \sin 2x.$

576. $y = (C_1 + C_2x + C_3x^2)e^x - \frac{1}{8}e^x \sin 2x.$

577. $y = C_1e^{-2x} + (C_2 \cos x + C_3 \sin x)e^x + \frac{1}{8}(2x^2 + 2x + 1)$

$+ \frac{1}{40}(\sin 2x + 3 \cos 2x) + \frac{1}{20} xe^x(3 \sin x - \cos x).$

578. $y = e^{2x} - e^{3x} + xe^{-x}.$

579. $y = -\frac{1}{3}(\cos 3x + \sin 3x - e^{3x}).$

580. $y = e^{2x}(\cos x - 2 \sin x) + (x + 1)^2e^x.$

581. $y = (x + \frac{3}{5})e^{-3x} + \frac{1}{5}(4 \sin x - 3 \cos x).$

582. $y = \cos x + x \sin x.$

583. $y = \cos 2x + \frac{1}{3}(\sin 2x + \sin x).$

584. $y = (1 - 3x)e^{3x} + \frac{x^2}{9} + \frac{x}{27} + \frac{1}{3}.$

585. $y = (4x + 2)e^{2x} + \frac{1}{2}x^2e^{2x}.$

586. $y = 3\pi \cos 2x + \frac{1}{2} \sin 2x + x(\sin 2x - \cos 2x).$

587. $y = e^{-x}(\sin x - 2 \cos x).$

588. $y = xe^x(\sin x - \cos x).$

589. $y = e^x - e^{-x} + x^2.$

590. $y = \cos x + 2 \sin x + e^{-x} - 3e^x + 2xe^x.$

591. $y = x^2.$

592. $y = 2xe^x.$

593. $y = \frac{1}{4} \cos x.$

594. $y = \sin 2x.$

595. $y = -1.$

596. $y = \cos x.$

597. $y = e^{-x}.$

598. $y = 3 + e^x.$

599. $y = -\frac{1}{5}.$

600. $y = e^x(\sin x + \cos x).$

601. $y = e^{-2x} \cos 2x.$

602. $y = (x^2 + x)e^{-x}.$

603. $y = C_1x + \frac{C_2}{x}.$

604. $y = \frac{1}{x}(C_1 + C_2 \ln x).$

605. $y = \frac{1}{\sqrt{x}}\left[C_1 \cos\left(\frac{\sqrt{23}}{2}\ln x\right) + C_2 \sin\left(\frac{\sqrt{23}}{2}\ln x\right)\right].$

606. $y = C_1 + C_2 \ln x.$

607. $y = \frac{C_1}{(x-2)^3} + C_2(x-2).$

608. $y = C_1(2x+1) + C_2(2x+1)\ln(2x+1).$

609. $y = C_1 + C_2x^2 + C_3x^4.$

610. $y = C_1 + C_2 \ln x + C_3x^3.$

611. $y = C_1 + \frac{C_2}{(x+1)^2} + C_3(x+1)^5.$

612. $y = C_1 + (2x+1)\left[C_2 \cos\frac{\ln(2x+1)}{\sqrt{2}} + C_3 \sin\frac{\ln(2x+1)}{\sqrt{2}}\right].$

613. $y = C_1x + C_2x^2 + C_3x^3.$

614. $y = C_1(x^3 - x) + C_2\left[6x^2 - 4 - 3(x^3 - x)\ln\left|\frac{x+1}{x-1}\right|\right].$

615. $y = C_1e^{-2x} + C_2(4x^2 + 1).$

616. $y = \frac{C_1}{x^2} + C_2(2x - 3).$

617. $y = C_1x^3 + C_2(x+1) - x.$

618. $y = C_1 \ln x + C_2x.$

619. $y = e^{x^2}(C_1e^x + C_2e^{-x}).$

620. $y = C_1 \cos(\sin x) + C_2 \sin(\sin x).$

621. $y = 1 + C_1x + C_2\sqrt{1 + x^2}.$

622. $y = C_1x^3 + \frac{C_2}{x} + x^4.$

623. $y = C_1(2x - 1) + \frac{C_2}{x} + x^2.$

624. $y = x + C_1 \cos e^x + C_2 \sin e^x.$

625. $y = C_1 + C_2 \sin x + \sin x \,.\, \ln|\sin x|.$

626. $y = \frac{C_1 + C_2\ln(x+1) + \ln^3(x+1)}{x+1}.$

627. $y = x^3 + C_1x^2 + C_2(2x - 1).$

628. $\frac{d^2x}{dt^2} = \frac{k}{m}x$ (x is the length of the overhanging part of the chain),

$t = \sqrt{\frac{6}{g}} \ln(6 + \sqrt{35})$ sec; $k = g$, $m = 6$.

629. $\frac{d^2s}{dt^2} = 1{\cdot}2t$, $s = 0{\cdot}2t^3 - t$.

630. $m\frac{d^2s}{dt^2} = -km$, $s = \frac{v_0^2}{2k}$.

631. $\frac{d^2x}{dt^2} = k^2x$, $x = ae^{-kt}$.

632. $y = C_1 \cos 2x + C_2 \sin 2x + \frac{1}{4}\cos 2x \ln|\cos 2x| + \frac{x}{2}\sin 2x$.

633. $y = C_1 \cos x + C_2 \sin x + \sin x \,.\, \ln\left|\tan\left(\frac{\pi}{4} + \frac{x}{2}\right)\right| - 2$.

634. $y = C_1e^x + C_2e^{-x} - 1 - xe^{-x} + (e^x - e^{-x})\ln(1 - e^{-x})$.

635. $y = C_1e^x + C_2 + (e^x + 1)\ln(1 + e^{-x})$.

636. $y = C_1 \cos x + C_2 \sin x + \frac{4}{3}\cos x \,.\, \sqrt{\cot x}$.

637. $y = C_1 \cos x + C_2 \sin x - \sqrt{\cos 2x}$.

638. $y = \frac{1}{x} + C_1e^x + C_2e^{-x} + C_3e^{2x}$.

639. $y = C_1 \cos x + C_2 \sin x + \frac{9}{4}\cos x \sqrt[3]{\cot x} + \frac{9}{10}\sin x \sqrt[3]{\tan^2 x}$.

640. $y = e^x(C_1 + C_2x - \ln\sqrt{x^2 + 1} + x \arctan x)$.

641. $y = (C_1 - x)e^{-x}\cos x + (C_2 + \ln \sin x)e^{-x}\sin x$.

642. $y = C_1e^x + C_2 - \cos e^x$.

643. $y = C_1 + C_2e^{-x} + e^{-x}\int \frac{e^x}{x}dx - \ln|x|$.

644. $y = C_1e^{-x} + C_2e^{-2x} + e^{-2x}\int \frac{e^{2x}\,dx}{x + 1}$.

645. $y = C_1 \cos x + C_2 \sin x - \cos x \int \frac{\cos x}{x}dx - \sin x \int \frac{\sin x}{x}dx$.

646. $y = C_1e^{x^2} + C_2 + (x^2 - 1)e^{x^2}$.

647. $y = C_1 + C_2 \tan x + \frac{1}{2}(1 + x \tan x)$.

648. $y = C_1(\ln x - 1)x + C_2 + x(\ln^2 x - 2\ln x + 2)$.

649. $y = C_1(x + \frac{1}{2})e^{-2x} + C_2 - x^2 + \frac{1}{2}$.

650. $y = C_1 x + C_2 e^x + \left(\frac{x^2}{2} - x\right) e^x$.

651. $y = C_1 \cos e^{-x} + C_2 \sin e^{-x} + e^{-x}$.

652. $y = C_1 e^{\frac{1}{x}} + C_2 \frac{1}{x} + 1 - \frac{\ln |x|}{x}$.

653. $y = C_1 e^x + C_2 \sin x + e^x(\cos x + \sin x)$.

654. $y = C_1 e^{x^2} + C_2 e^{-x^2} + (2x^2 - 1)e^{x^2}$.

655. $y = C_1 e^x + C_2 \ln x + e^x(x - x \ln x + \ln x)$.

656. $y = \frac{2}{3}\sqrt{x^2 + x}$.

657. $y = \frac{x + 1}{\cos x}$.

658. $y = -\frac{\cos 2x}{2 \sin x}$.

659. $y = 1$.

660. $y = \frac{1}{x}$.

661. $y = \frac{1}{2}(\arctan x)^2$.

662. $y = \left(1 - x + \frac{x^2}{2}\right) e^x$.

663. $y = \frac{1 - \ln x}{\sqrt{x}}$.

664. $y = (x - 1)e^x$.

665. $y = \frac{1}{x}$.

666. $y = 1$.

677. (*a*) $\lambda = k^2, \quad k = 0, 1, 2, \ldots$

(*b*) $\lambda = 4k^2, \quad k = 0, 1, 2, \ldots$

678. For any λ.

679. (*a*) Soluble $y = \frac{\operatorname{sh} x}{\operatorname{sh} 2\pi}$.

(*b*) Insoluble.

680. (1) $\lambda - \omega^2 > 0, \quad y = C_1 \cos 2n\pi x + C_2 \sin 2n\pi x;$
(2) $\lambda - \omega^2 = 0, \quad y \equiv C;$
(3) $\lambda - \omega^2 < 0, \quad y \equiv 0.$

681. $y = +\sqrt{5 - (x-2)^2}.$

682. $$y(x) = \frac{e^{-\alpha s x} - e^{-\alpha s(2x_0 - x)}}{1 - e^{-2\alpha s x_0}} v_1 + \frac{e^{-\alpha s(x_0 - x)} - e^{-\alpha s(x_0 + x)}}{1 - e^{-2\alpha s x_0}} v_2.$$

683. $$y(x) = \frac{e^{-\alpha x\sqrt{s}} + e^{-\alpha(2x_0 - x)\sqrt{s}}}{1 + e^{-2\alpha x_0\sqrt{s}}} v.$$

684. (*a*) $$y = \frac{1}{s}\frac{\operatorname{ch} \alpha\sqrt{s}(x - x_0)}{\operatorname{ch} \alpha\sqrt{s}x_0};$$

(*b*) $$y = -\frac{1}{\alpha s\sqrt{s}}\frac{\operatorname{sh} \alpha\sqrt{s}(x - x_0)}{\operatorname{ch} \alpha\sqrt{s}x_0}.$$

685. $$y = \frac{gl}{s^2}\left[\frac{e^{\alpha s x} + e^{\alpha s(x_0 - x)}}{1 + e^{\alpha s x_0}} - 1\right].$$

686. (*a*) $$y = \frac{1}{\lambda}\frac{\operatorname{sh} \lambda x}{\operatorname{sh} \lambda}.$$

(*b*) $$y = \frac{1}{\lambda}\frac{\operatorname{ch} \lambda x}{\operatorname{ch} \lambda}.$$

(*c*) $$y = \frac{1}{\lambda^2}\frac{\operatorname{sh} \lambda x}{\operatorname{ch} \lambda}.$$

(*d*) $$y = \frac{1}{\lambda^2}\frac{\operatorname{ch} \lambda x}{\operatorname{sh} \lambda}.$$

687. $y = C \sin \lambda x, \quad \lambda = 0, 1, 2, \ldots$

688. $y = C[e^{\lambda x} - e^{\lambda(2\pi - x)} - (1 - e^{2\lambda\pi}) \cos \lambda x - (1 + e^{2\lambda\pi}) \sin \lambda x],$
$\operatorname{th} \lambda\pi = \tan \lambda\pi.$

689. $$y = C\left[\frac{(\cos \lambda\pi - \sin \lambda\pi - e^{-\lambda\pi})e^{\lambda x} + (e^{\lambda\pi} - \cos \lambda\pi - \sin \lambda\pi)e^{-\lambda x}}{2(\operatorname{ch} \lambda\pi - \cos \lambda\pi)}\right.$$
$$\left. + \frac{2(\sin \lambda\pi - \operatorname{sh} \lambda\pi) \cos \lambda x + 2(\operatorname{ch} \lambda\pi - \cos \lambda\pi) \sin \lambda x}{2(\operatorname{ch} \lambda\pi - \cos \lambda\pi)}\right],$$
$\operatorname{ch} \lambda\pi \cos \lambda\pi = 1.$

690. $$y = \frac{x - 1}{\alpha} e^{-\alpha x}.$$

691. $$y = \frac{2}{75} \sin x + \frac{4}{75} \sin 2x + \frac{1}{25} \cos 2x.$$

692. $y = 0.$

693. $y = C(x \ln x - x + 1)$.

694. $y = 0$.

695. $y = 1 + x^2 + \frac{x^4}{2!} + \frac{x^6}{3!} + \ldots \quad (= e^{x^2})$.

696. $y = C_1\left(1 - \frac{x}{2!} + \frac{x^2}{4!} - \ldots\right) + C_2\sqrt{x}\left(1 - \frac{x}{3!} + \frac{x^2}{5!} - \ldots\right);$

$(y = C_1 \cos\sqrt{x} + C_2 \sin\sqrt{x})$.

697. $y = C \sum_{k=1}^{\infty} \frac{n(n-1)\ldots(n-k+1)}{k!} x^k$.

698. $y = C_1\left(1 + \frac{x}{3} + \frac{1\,.\,4x^2}{3\,.\,6} + \frac{1\,.\,4\,.\,7x^3}{3\,.\,6\,.\,9} + \ldots\right)$

$+ C_2 x^{\frac{7}{3}}\left(1 + \frac{8x}{10} + \frac{8\,.\,11x^2}{10\,.\,13} + \frac{8\,.\,11\,.\,14x^3}{10\,.\,13\,.\,16} + \ldots\right)$.

699. $y = C_1\left[1 - \frac{x^2}{2} + \frac{x^4}{8} - \ldots + \frac{(-1)^n x^{2n}}{2\,.\,4\ldots 2n} + \ldots\right]$

$+ C_2\left[x - \frac{x^3}{1\,.\,3} + \frac{x^5}{1\,.\,3\,.\,5} - \frac{x^7}{1\,.\,3\,.\,5\,.\,7} + \ldots\right.$

$\left.+ \frac{(-1)^{n-1}x^{2n-1}}{1\,.\,3\,.\,5\ldots(2n-1)} + \ldots\right]$.

700. $y = \frac{x^2}{2} + \frac{x^4}{4!} + \frac{3x^6}{6!} + \frac{3\,.\,5x^8}{8!} + \ldots + \frac{(2n-1)!!x^{2n+2}}{(2n+2)!} + \ldots$

$(2n-1)!! = 1\,.\,3\,.\,5\ldots(2n-1)$.

701. $y = -2 + 2x - x^2 + \frac{x^3}{3} - \frac{x^4}{4} + \frac{7x^5}{60} - \ldots$

702. $y = 1 + \frac{2x^4}{4!} - \frac{2x^5}{5!} + \frac{2x^6}{6!} - \frac{2x^7}{7!} + \frac{62x^8}{8!} - \ldots$

703. $y = C_1\left(1 + \frac{x^2}{2} + \frac{x^3}{6} + \frac{x^4}{12} + \frac{x^5}{24} + \frac{13x^6}{720} + \ldots\right)$

$+ C_2\left(x + \frac{x^3}{6} + \frac{x^4}{12} + \frac{x^5}{30} + \frac{x^6}{72} + \frac{29x^7}{5{,}040} + \ldots\right)$.

704. $y = \frac{x^3}{3} + \frac{x^7}{63} + \frac{2x^{11}}{3\,.\,11\,.\,63} + \frac{13x^{15}}{5\,.\,11\,.\,63^2} + \frac{46x^{19}}{3\,.\,5\,.\,11\,.\,19\,.\,63^2}$

$+ \frac{15{,}178x^{23}}{5\,.\,11^2\,.\,19\,.\,23\,.\,63^3}$.

705. $y = x + \frac{x^2}{2} + \frac{2x^3}{3} + \frac{11x^4}{24} + \frac{53x^5}{120} + \frac{767x^6}{2{,}160} + \dots$

706. $y = C_1 J_{\frac{1}{3}}(2x) + C_2 J_{-\frac{1}{3}}(2x).$

707. $y = C_1 J_{\frac{1}{2}}(x) + C_2 J_{-\frac{1}{2}}(x).$

708. $y = C_1 J_0\left(\frac{x}{3}\right) + C_2 Y_0\left(\frac{x}{3}\right).$

709. $y = C_1 J_0(2x) + C_2 Y_0(2x).$

710. $y = x^{\frac{3}{2}}\left[C_1 J_{\frac{5}{4}}(x^2) + C_2 J_{-\frac{5}{4}}(x^2)\right].$

711. $y = \sqrt[4]{x}\left[C_1 J_{\frac{1}{2}}(\sqrt{x}) + C_2 J_{-\frac{1}{2}}(\sqrt{x})\right].$

712. $y = \frac{1}{x^2}\left[C_1 J_2(x) + C_2 Y_2(x)\right].$

713. $y = \frac{1}{x}\left[C_1 J_1(2x) + C_2 Y_1(2x)\right].$

724. $\left.\begin{aligned} x &= t + C_1 \cos 2t + C_2 \sin 2t \\ y &= 1 + C_1 \sin 2t - C_2 \cos 2t \end{aligned}\right\}.$

725. $\left.\begin{aligned} x &= (C_1 - C_2)e^{2t} \cos t - (C_1 + C_2)e^{2t} \sin t \\ y &= (C_1 \sin t + C_2 \cos t)e^{2t} \end{aligned}\right\}.$

726. $\left.\begin{aligned} x &= e^{-2t}(1 - 2t) \\ y &= e^{-2t}(1 + 2t) \end{aligned}\right\}.$

727. $\left.\begin{aligned} x &= t^2 + t + C_1 e^{2t} + C_2 e^{3t} \\ y &= t + 1 + 2C_1 e^{2t} \end{aligned}\right\}.$

728. $\left.\begin{aligned} x &= e^{-6t}(C_1 \cos t + C_2 \sin t) \\ y &= e^{-6t}[(C_1 + C_2) \cos t - (C_1 - C_2) \sin t] \end{aligned}\right\}.$

729. $\left.\begin{aligned} x &= (C_1 - 3C_1 t - 3C_2)e^{5t} \\ y &= (C_1 t + C_2)e^{5t} \end{aligned}\right\}.$

730. $\left.\begin{aligned} x &= C_1 e^{-t} + C_2 e^{2t} \\ y &= C_3 e^{-t} + C_2 e^{2t} \\ z &= -(C_1 + C_3)e^{-t} + C_2 e^{2t} \end{aligned}\right\}.$

731. $\left.\begin{aligned} x &= C_1 e^{3t} + C_2 e^{-2t} \\ y &= \frac{3}{2} C_1 e^{3t} - C_2 e^{-2t} + C_3 e^{-t} \\ z &= \frac{3}{2} C_1 e^{3t} - C_2 e^{-2t} - C_3 e^{-t} \end{aligned}\right\}.$

732. $\left.\begin{aligned} x &= C_1e^{-2t} + C_2 \sin 4t + C_3 \cos 4t \\ y &= -\tfrac{1}{4}C_1e^{-2t} + \tfrac{1}{2}C_2 \cos 4t - \tfrac{1}{2}C_3 \sin 4t \\ z &= -\tfrac{1}{4}C_1e^{-2t} + C_2 \sin 4t + C_3 \cos 4t \end{aligned}\right\}.$

733. $\left.\begin{aligned} x &= C_1e^{t} + C_2 \sin t + C_3 \cos t \\ y &= t - C_1e^{t} + C_2 \cos t - C_3 \sin t \\ z &= 1 + C_2 \sin t + C_3 \cos t \end{aligned}\right\}.$

734. $\left.\begin{aligned} x + y - z &= C_1e^{-t} + \tfrac{1}{2}e^{t} + \tfrac{1}{4}e^{3t} - 4 \\ x + y + 2z &= C_2e^{2t} - e^{t} + e^{3t} - 4 \\ x - y &= C_3e^{-2t} + \tfrac{1}{3}e^{t} - \tfrac{1}{5}e^{3t} \end{aligned}\right\}.$

735. $\left.\begin{aligned} x &= C_1e^{\sin t} \\ y &= C_2e^{\operatorname{sh} t} \end{aligned}\right\}.$

736. $\left.\begin{aligned} x &= \frac{4}{25}e^{t} - \frac{1}{36}e^{2t} \\ y &= \frac{1}{25}e^{t} + \frac{7}{36}e^{2t} \end{aligned}\right\}.$

737. $\left.\begin{aligned} x &= 2(2e^{t} + e^{-t}) \\ y &= -e^{t} - e^{-t} \end{aligned}\right\}.$

738. $\left.\begin{aligned} x &= \cos t + \sin t \\ y &= \cos t - \sin t \end{aligned}\right\}.$

739. $\left.\begin{aligned} x &= (1 - 2t)e^{-2t} \\ y &= te^{-2t} \end{aligned}\right\}.$

740. $\left.\begin{aligned} x &= -5e^{2t} \sin t \\ y &= e^{2t}(\cos t - 2 \sin t) \end{aligned}\right\}.$

741. $\left.\begin{aligned} x &= -\frac{4}{3}t - \frac{7}{9} \\ y &= \frac{1}{3}t - \frac{5}{9} \end{aligned}\right\}.$

742. $\left.\begin{aligned} x &= (\sin t - 2 \cos t)e^{-t} \\ y &= e^{-t} \cos t \end{aligned}\right\}.$

743. $\left.\begin{aligned} x &= e^{5t} + e^{3t} \\ y &= 6e^{5t} - 7e^{3t} \end{aligned}\right\}.$

744. $\left.\begin{aligned} x &= \cos t \\ y &= -\sin t \end{aligned}\right\}.$

745. $\left.\begin{array}{l} x = -(2t + \sin t + \cos t + e^{-t}) \\ y = \cos t - 2e^{-t} + 2 \end{array}\right\}.$

746. $\left.\begin{array}{l} x = e^{2t} + e^{3t} + t^2 + t \\ y = 2e^{2t} + t + 1 \end{array}\right\}.$

747. $\left.\begin{array}{l} x = \sqrt{\dfrac{2t + C_2}{C_1}} \\ y = \sqrt{C_1(2t + C_2)} \end{array}\right\}.$

748. $\left.\begin{array}{l} lt + mx + ny = C_1 \\ t^2 + x^2 + y^2 = C_2 \end{array}\right\}.$

749. $\left.\begin{array}{l} x^2 + y^2 = C_1{}^2 \\ p^2 + q^2 = C_2{}^2 \\ xp + yq = C_3 \end{array}\right\}.$

750. $\left.\begin{array}{l} x^2 + y^2 + t^2 = C_1 \\ x^2 - 2xy - y^2 = C_2 \end{array}\right\}.$

751. $\left.\begin{array}{l} x = \dfrac{1}{3} t + C_2 \dfrac{1}{t^2} \\ y = C_1 e^t - \dfrac{t}{3} - \dfrac{C_2}{t^2} \end{array}\right\}.$

752. $\left.\begin{array}{l} x^2 + y^2 = C_1 x - t^2 \\ y = C_2 x \end{array}\right\}.$

753. $\left.\begin{array}{l} x = C_1 t. \\ \ln\left| y - \dfrac{2t}{x} \right| = t + C_2 \end{array}\right\}.$

754. $\left.\begin{array}{l} t^2 - x^2 = C_1 \\ x^2 - y^2 = C_2 \end{array}\right\}.$

755. Unstable.

756. Unstable.

757. Stable.

758. Stable.

759. Stable.

760. Stable.

761. Unstable focus.

762. Saddle point.

763. Centre.

764. Stable focus.

765. Unstable node.

766. Stable node.

767. Unstable node.

768. Stable node.

769. Unstable node.

770. $\alpha < -\frac{3}{2}$.

771. Unstable.

772. Stable.

773. Stable.

774. Stable.

775. Stable.

776. Unstable.

777. Stable.

778. Unstable.

779. Stable.

780. Stable.

781. Unstable.

782. Unstable.

783. Stable.

784. Stable.

785. Unstable.

786. Stable.

787. Unstable.

788. Stable.

789. Stable.

790. Stable.

791. Unstable.

792. Stable.

793. $\alpha > \frac{3}{2}$.

794. Always unstable.

795. $\alpha > \frac{5}{2}$.

796. $\alpha > 0, \quad \alpha\beta > 1$.

797. $\alpha > \frac{2}{3}, \quad \beta > 0, \quad 9\beta - 6\alpha + 4 < 0$.

798. Stable.

799. Stable.

800. Stable.

801. Stable.

802. Stable.

803. Stable.

804. Stable.

805. Unstable.

806. Unstable.

807. Unstable.

808. Unstable.

809. Stable.

810. Unstable.

811. Stable.

812. Stable.

813. Unstable.

814. Unstable.

815. Unstable.

816. $\frac{2}{p^3} - \frac{2}{p^2} + \frac{2}{p}$.

817. $\frac{6}{p^4} + \frac{8}{p^3} + \frac{4}{p^2}$.

818. $\frac{6e^{-2p}}{p^4}$.

819. $\frac{1}{(p + \alpha)^2}$.

820. $\frac{2p}{(p - 1)^3}$.

821. $\frac{p^2 - 2a^2}{p(p^2 - 4a^2)}$.

822. $e\frac{p^2 + 1}{(p + 1)^3}$.

823. $\frac{\beta}{(p - \alpha)^2 + \beta^2}$.

824. $\frac{1}{2}\left[\frac{p - 3}{(p - 3)^2 + 49} + \frac{p - 3}{(p - 3)^2 + 1}\right]$.

825. $\frac{1}{(p - \lambda)^2 + 1} e^{-\alpha p}$.

826. $\frac{1}{\sqrt{2}} \frac{p - 1}{p^2 - 4p + 5}$.

827. $\dfrac{(p-\alpha)\cos\beta - \sin\beta}{(p-\alpha)^2+1}$.

828. $\arctan\dfrac{1}{p}$.

829. $\arctan\dfrac{1}{p+\lambda}$.

830. $\dfrac{20p}{p^4+58p^2+441}$.

831. $\dfrac{8}{p(p^2+16)}$.

832. $\dfrac{1}{p+2-i}$.

833. $\dfrac{1}{p-\alpha-i\beta}$.

834. $\dfrac{p^3+20p}{p^4+40p^2+144}$.

835. $\dfrac{p^2+32}{p^3+64p}$.

836. $\frac{3}{5}+\frac{1}{5}(4\sin t-3\cos t)e^{-2t}$.

837. $t \operatorname{ch} at$.

838. t^k.

839. $e^{3t}-e^t$.

840. $\dfrac{3}{2}e^{-2t}\cos\sqrt{\dfrac{11}{2}}\,t+\dfrac{13}{\sqrt{22}}e^{-2t}\sin\sqrt{\dfrac{11}{2}}\,t$.

841. $\dfrac{4\sqrt{3}}{9}\,e^{-\frac{t}{2}}\sin\dfrac{\sqrt{3}}{2}\,t-\dfrac{2}{3}\,te^{-\frac{t}{2}}\cos\dfrac{\sqrt{3}}{2}\,t$.

842. $\dfrac{1}{9}(e^{-2t}-e^t+3te^t)$.

843. $(\sqrt{2}-1)e^t+(\sqrt{2}+1)e^{-t}-2\sqrt{2}e^{-\sqrt{2}t}$.

844. $\dfrac{2}{\sqrt{3}}\,e^{-\frac{t}{2}}\sin\dfrac{\sqrt{3}}{2}\,t$.

845. $\frac{1}{2}(\operatorname{sh} t-\sin t)$.

846. $\cos 2(t-2)\,.\,\eta(t-2)$.

847. $\frac{1}{3}\sin 3(t-\frac{1}{2})\,.\,\eta(t-\frac{1}{2})$.

848. $3t^2e^{-t} + te^{-2t}$.

849. $2e^{3t}\cos\sqrt{3}t + \dfrac{11}{\sqrt{3}}e^{3t}\sin\sqrt{3}t$.

850. $\dfrac{1}{4}\left(\operatorname{sh} t - \dfrac{\sqrt{3}}{3}\sin\sqrt{3}t\right)$.

851. $\left(t - \dfrac{3}{2}\right) \cdot \eta\left(t - \dfrac{3}{2}\right)$.

852. $x = e^{-2t} - e^{-3t}$.

853. $x = -(t^3 + 2t^2 + 2t + 1)$.

854. $x = \sin t$.

855. $x = e^{-t} + \sin t - \cos t$.

856. $x = \dfrac{t^2 - 2}{4}e^{-3t}$.

857. $x = \frac{1}{3} + 3e^{-t} - \frac{1}{3}e^{-3t}$.

858. $x = \frac{1}{2}$.

859. $x = 2$.

860. $x = \frac{1}{3}t$.

861. $x = \dfrac{1 + t}{2}$.

862. $x = t^2 - 3t + 4$.

863. $x = -(t^2 + t + 1)$.

864. $x = 1 + t + t^2 + e^{7t}$.

865. $x = t^3 - \dfrac{3}{2}t$.

866. $x = \dfrac{3t^2 - t}{36}$.

867. $x = \sin t + e^t$.

868. $x = 2 - \dfrac{t(2t + 1)}{56}e^{-2t}$.

869. $x = \left(\dfrac{t^3}{6} - \dfrac{t^2}{2} + t\right)e^{2t}$.

870. $x = \left(\dfrac{t^2}{8} + t - 2\right)e^{\frac{t}{2}}$.

871. $x = (1 + t)e^{-t} + (1 - t)e^{-2t}$.

872. $x = \frac{t}{5}(6e^{3t} - 2e^{-2t})$.

873. $x = \frac{t^4}{12}e^{-2t}$.

874. $x = e^t + \cos t - \sin t$.

875. $x = 3(1 + t)\sin 3t$.

876. $x = t(\sin 2t + \frac{1}{8}\cos 2t)$.

877. $x = \frac{t-1}{4}(\cos t + \sin t)$.

878. $37x = (36e^t + 1)\cos 3t - 6\sin 3t$.

879. $x = e^{2t}[(1 - t)\cos t + (1 + t)\sin t]$.

880. $x = t - 1 + 2e^t$.

881. $x = e^t + e^{-\frac{t}{2}}\left(\cos\frac{\sqrt{3}}{2}t + \sin\frac{\sqrt{3}}{2}t\right) - 2$.

882. $x = e^t + e^{-t}(\cos t + \sin t) + \left(\frac{t^2}{10} - \frac{4t}{25}\right)e^t$.

883. $x = \frac{24}{17}\sin t - \frac{40}{17}\cos t$.

884. $x = 2\sqrt{2}t\sin\left(2t - \frac{\pi}{4}\right)$.

885. $\left.\begin{array}{l} x = e^t + e^{-t} \\ y = -e^t + e^{-t} \end{array}\right\}$.

886. $\left.\begin{array}{l} x = 4e^{-2t} - 3e^{-3t} \\ y = 3e^{-3t} - 2e^{-2t} \end{array}\right\}$.

887. $\left.\begin{array}{l} x = e^t(\cos t - 2\sin t) \\ y = e^t(3\sin t + \cos t) \end{array}\right\}$.

888. $\left.\begin{array}{l} x = -\frac{5}{4} + \frac{13}{4}\cos 2t - 3\sin 2t \\ y = \frac{3}{2}t + 3\cos 2t + \frac{13}{4}\sin 2t \end{array}\right\}$.

889. $\left.\begin{array}{l} x = e^t \\ y = e^t \end{array}\right\}$.

890. $$\left.\begin{aligned} x &= -\frac{1}{2} + e^t - \frac{11}{34}e^{4t} - \frac{3}{17}\cos t + \frac{5}{17}\sin t \\ y &= -\frac{2}{3}e^t + \frac{22}{51}e^{4t} + \frac{4}{17}\cos t - \frac{1}{17}\sin t \end{aligned}\right\}.$$

891. $$\left.\begin{aligned} x &= 2 - e^t \\ y &= -2 + 4e^t - te^t \\ z &= -2 + 5e^t - te^t \end{aligned}\right\}.$$

892. $$\left.\begin{aligned} x &= 1 + 3e^{2t} + e^{-2t} \\ y &= e^{2t} - e^{-2t} \\ z &= 2e^{2t} + 2e^{-2t} \end{aligned}\right\}.$$

893. $$\left.\begin{aligned} x &= 1 + 4te^{-t} \\ y &= e^{-t} \\ z &= -1 - 3e^{-t} \end{aligned}\right\}.$$

894. $$\left.\begin{aligned} x &= 2(1 - e^{-t} - te^{-t}) \\ y &= 2 - t - 2e^{-t} - 2te^{-t} \end{aligned}\right\}.$$

895. $$\left.\begin{aligned} x &= e^t + \sin t \\ y &= e^t - \sin t \end{aligned}\right\}.$$

896. $$\left.\begin{aligned} x &= \cos t + e^{-\sqrt{3}t} \\ y &= \tfrac{1}{2}(\cos t - e^{-\sqrt{3}t}) \end{aligned}\right\}.$$

897. $$\left.\begin{aligned} x &= t - \frac{t^3}{6} + e^t \\ y &= 1 + \frac{1}{24}t^4 - e^t \end{aligned}\right\}.$$

898. $$\left.\begin{aligned} x &= 12(\operatorname{ch} t - 1) - \frac{7}{2}t \operatorname{sh} t \\ y &= 7t \operatorname{sh} t - 17(\operatorname{ch} t - 1) \end{aligned}\right\}.$$

899. $$\left.\begin{aligned} x &= t^2 + t \\ y &= -\tfrac{1}{2}t^2 \end{aligned}\right\}.$$

900. $$\left.\begin{aligned} x &= e^t(2\cos t - \sin t) \\ y &= e^t(3\cos t + \sin t) \end{aligned}\right\}.$$